Mario Pitschmann

On the Quantisation of Topological Field Models

Mario Pitschmann

On the Quantisation of Topological Field Models

The Thirring Model and Sine-Gordon Model

Südwestdeutscher Verlag für Hochschulschriften

Impressum/Imprint (nur für Deutschland/ only for Germany)
Bibliografische Information der Deutschen Nationalbibliothek: Die Deutsche Nationalbibliothek verzeichnet diese Publikation in der Deutschen Nationalbibliografie; detaillierte bibliografische Daten sind im Internet über http://dnb.d-nb.de abrufbar.
Alle in diesem Buch genannten Marken und Produktnamen unterliegen warenzeichen-, marken- oder patentrechtlichem Schutz bzw. sind Warenzeichen oder eingetragene Warenzeichen der jeweiligen Inhaber. Die Wiedergabe von Marken, Produktnamen, Gebrauchsnamen, Handelsnamen, Warenbezeichnungen u.s.w. in diesem Werk berechtigt auch ohne besondere Kennzeichnung nicht zu der Annahme, dass solche Namen im Sinne der Warenzeichen- und Markenschutzgesetzgebung als frei zu betrachten wären und daher von jedermann benutzt werden dürften.

Verlag: Südwestdeutscher Verlag für Hochschulschriften Aktiengesellschaft & Co. KG
Dudweiler Landstr. 99, 66123 Saarbrücken, Deutschland
Telefon +49 681 37 20 271-1, Telefax +49 681 37 20 271-0, Email: info@svh-verlag.de
Zugl.: Wien, TU, Diss., 2006

Herstellung in Deutschland:
Schaltungsdienst Lange o.H.G., Berlin
Books on Demand GmbH, Norderstedt
Reha GmbH, Saarbrücken
Amazon Distribution GmbH, Leipzig
ISBN: 978-3-8381-0276-4

Imprint (only for USA, GB)
Bibliographic information published by the Deutsche Nationalbibliothek: The Deutsche Nationalbibliothek lists this publication in the Deutsche Nationalbibliografie; detailed bibliographic data are available in the Internet at http://dnb.d-nb.de.
Any brand names and product names mentioned in this book are subject to trademark, brand or patent protection and are trademarks or registered trademarks of their respective holders. The use of brand names, product names, common names, trade names, product descriptions etc. even without a particular marking in this works is in no way to be construed to mean that such names may be regarded as unrestricted in respect of trademark and brand protection legislation and could thus be used by anyone.

Publisher:
Südwestdeutscher Verlag für Hochschulschriften Aktiengesellschaft & Co. KG
Dudweiler Landstr. 99, 66123 Saarbrücken, Germany
Phone +49 681 37 20 271-1, Fax +49 681 37 20 271-0, Email: info@svh-verlag.de

Copyright © 2009 by the author and Südwestdeutscher Verlag für Hochschulschriften Aktiengesellschaft & Co. KG and licensors
All rights reserved. Saarbrücken 2009

Printed in the U.S.A.
Printed in the U.K. by (see last page)
ISBN: 978-3-8381-0276-4

Contents

Preface   iii

1. Thirring Model   1
   1.1. Introduction 1
   1.2. Generating functional of Green functions 3
   1.3. Two–point causal Green function $G(x,y)$ 5
   1.4. Two–point correlation function $C(x,y)$ 6
   1.5. Non–perturbative renormalisation of the massless Thirring model 7
   1.6. The Schwinger term 9
   1.7. The norm of the wave functions of the states related to the components of the vector current 10
      1.7.1. The free theory 10
      1.7.2. The interacting theory 11
   1.8. Conclusion 13

2. Sine–Gordon Model   15
   2.1. Introduction to the "classical" sine–Gordon model 15
   2.2. Introduction to the "quantum" sine–Gordon model 23
   2.3. Power counting and renormalisation of the sine–Gordon model 26
   2.4. Renormalisation of the causal two–point Green function 28
      2.4.1. Expansion to first order in α and to all orders in β^2 29
      2.4.2. Expansion to second order in α and to all orders in β^2 31
   2.5. Renormalisation of the massive sine–Gordon model 32
   2.6. Renormalisation of the massless sine–Gordon model around solitons 34
   2.7. Renormalisation of the soliton mass by Gaussian quantum corrections in continuous space–time 40
   2.8. Renormalisation of the soliton mass by Gaussian quantum corrections in discretised space–time 42
      2.8.1. Periodic boundary conditions 44
      2.8.2. Anti-periodic boundary conditions 48
      2.8.3. Rigid walls 51
      2.8.4. Remark on the additional finite correction to the soliton mass 54
   2.9. Conclusion 58

A. Thirring Model   59

A.1. Generating functional of Green functions 59
    A.1.1. Evaluation of the Green function 59
    A.1.2. Evaluation and regularisation of the vector current 63
    A.1.3. Evaluation of the derivative of the vector current 65
    A.1.4. Performing the path integral 67
    A.1.5. Evaluation of $Z_{\text{Th}}^{(g)}[A;0,0]$ 67
A.2. Two–point causal Green function $G(x,y)$ 69
    A.2.1. Evaluation of the Green function 69
    A.2.2. Evaluation of $\Delta_F(0;\mu)$ 73
A.3. Two–point correlation function $C(x,y)$ 74
A.4. The Schwinger term for the free theory 75
A.5. The norm of the current states for the free theory 76
    A.5.1. The time-ordered current correlation function 76
    A.5.2. The norm of the states 76

B. Sine–Gordon Model   79

B.1. Expansion to second order in α 79
    B.1.1. The massless sine–Gordon model 79
    B.1.2. Momentum representation 83
    B.1.3. The massive sine–Gordon model 86
    B.1.4. The momentum representation 88
B.2. Exact solutions concerning Gaussian quantum corrections around a soliton . 89

Preface

In my thesis "On the Quantisation of Topological Field Models" some aspects of the quantum theory of topological field models were investigated. It consists of several parts dealing with the quantisation of topological quantum field theories in 1+1 and 3+1 space-time dimensions. Since my investigations in four space-time dimensions were less complete I have included in this book only those parts of my thesis dealing with quantum field theories in two space-time dimensions albeit in a slightly extended version. These quantum field theories are the Thirring model and the sine–Gordon model. Maybe there will be a second edition of this book containing further investigations in four space-time dimensions.

The Thirring model is a fermionic quantum field theory which is exactly solvable and hence an interesting laboratory to study quantum effects in a non-perturbative way. While it is well known that the massive Thirring model is renormalisable, the massless model was thought to be non-renormalisable. The reason for this lies in the fact that for renormalisability it is necessary that all dynamical dimensions calculated from different correlation functions are equal. Only in this case it is possible to remove all divergences by renormalisation of the fermion wave functions. The solutions of the massless Thirring model obtained so far do not allow to have equal dynamical dimensions for all correlation functions and hence it is not possible to renormalise the massless Thirring model with these solutions. In this book a more general solution of the massless Thirring model is obtained, which is parametrised by two parameters. For special values of these parameters all known solutions can be obtained. Using this generalised solution it is possible to have equal dynamical dimensions and to renormalise the massless Thirring model.

In the second part of this book the quantisation of the sine–Gordon model is investigated. This soliton model has a wide range of applications. It is exactly solvable and equivalent to the massive Thirring model. In this book it is shown that the divergences appearing in the sine–Gordon model can be removed by the renormalisation of the dimensional coupling constant $\alpha_0(\Lambda^2)$. Using path integral techniques the quantum correction to the mass of a soliton are calculated in continuous space–time and within the discretisation technique with periodic and anti-periodic boundary conditions and rigid walls. The obtained results agree completely. Finally, it is shown that the finite contribution to the quantum mass of a soliton found in the literature arises due to a non–covariant procedure.

Für meine Familie

Acknowledgements

First of all I would like to thank the supervisors of my thesis Prof. Manfried Faber and Prof. Andrei Ivanov for their support during the work on my thesis and this book. They had always time to discuss my ideas with them. The most important lesson I have learned from Prof. Faber is that there is often a simple geometrical picture which explains more than a lot of complicated equations, while Prof. Ivanov has taught me a lot on mathematics and has shown me how to apply it to physics with ease and beauty.

I would like to thank my long time colleagues Hidir Bozkaya, Roman Höllwieser, Gerald Jordan and Max Wellenzohn for their collaboration and friendship. They had always an open ear for my problems and we had many interesting discussions. Furthermore, I would like to thank Roman Bertle for helping me a lot on questions of computers and many interesting discussions.

Special thanks goes to my whole family, since they have always supported me and have helped me through all the minor and major problems which wait for every sailor who goes through the stormy sea of life.

"When all else fails, you can always tell the truth."

– Abdus Salam

On the Quantisation of Topological Field Models

The Thirring Model and Sine-Gordon Model

1. Thirring Model

1.1. Introduction

In 1958 Thirring invented the model of a self-interacting spinor field in 1+1 dimensions [1] and showed that this model is exactly soluble in the sense of the construction of the eigenstates of the Hamiltonian. Shortly afterwards, Glaser proposed a solution of the massless model by displaying the field operator as an explicit functional of the corresponding incoming field [2] (see also Refs. [3, 4]). He found that after an infinite field renormalisation all matrix elements of the field operators are finite analytical functions of the coupling constant. So far, the authors solving the massless Thirring dealt with formal manipulations. Glaser has made use of asymptotic fields but there is no dimensional parameter in the massless theory since the coupling constant is dimensionless. Johnson solved the model by calculating Green functions instead of seeking a formal operator solution [5]. Green functions exist even if there is no single-particle state in the theory (i.e. there is no asymptotic condition). Johnson used the equations expressing the conservation of the current and its dual to integrate directly the vertex function. In order to regularise the singular vector current operator Johnson used the point-splitting technique and was able to obtain a covariant limit and a consistent solution of the model. Sommerfield using the same regularisation of the current has calculated correlation functions by using external-source techniques [6]. In the limit of vanishing external sources the solutions of Sommerfield and Johnson agree themselves. Since the current regularisation used by them performs partly a limit in time-like direction they violate the usual requirement of a canonical theory which states that all operators should be defined solely on a given space-like surface. To overcome this problem Hagen has proposed a solution of the massless Thirring model by reducing the massless Thirring model to the massless Schwinger model with fermions coupled to an external vector field [7, 8]. Within such an approach Hagen calculated Green functions and found that the dynamical dimension of massless Thirring fermions are parametrised by one free parameter, which he denoted as ξ. Such a parameter does not appear in the Lagrangian and generates a class of solutions which incorporates the solutions of Johnson and Sommerfield as a special case. Independently, during the same period Klaiber [9] proposed a procedure for the calculation of correlation functions in the massless Thirring model using a canonical operator technique. He showed that this class of solutions is also parametrised by one free parameter. The procedure developed by Klaiber suffers from problems related to the use of the regularisation, which breaks the chiral symmetry of the massless Thirring model explicitly and produces at the intermediate steps of the calculation a non-vanishing divergence of the axial vector current due to non-covariant terms. Since Hagen's solution does not suffer from such a problem and reproduces Klaiber's results such a violation of chiral symmetry is not crucial. The most important problem of these solutions both Klaiber's and Hagen's lies in the different dynamical dimensions of massless Thirring fermions calculated from different correlation functions. Such a problem is related to the impossibility to remove the dependence of the correlation functions on the ultraviolet cut–off by means of the renormalisation of wave functions of Thirring fermions. In other words this testifies that the massless Thirring model is not renormalisable in the sense of Jackiw [11] in spite of the fact that it is solvable.

Due to Jackiw a quantum field theory in 1+1 dimensions is renormalisable in the sense that the dependence of the causal Green functions and correlation functions of left–right fermion densities on the ultraviolet cut–off can be removed by the renormalisation of the wave function of massless Thirring fermion fields, if the dynamical dimensions, calculated from different correlation functions, are equal [11]. For all known solutions of the massless Thirring model obtained so far [5]–[7], [9], [13]–[19] the dynamical dimensions of massless Thirring fermion fields, calculated from causal Green functions and left–right correlation functions, are different. According to Jackiw this implies that the massless Thirring model would be a non-renormalisable theory if there would not exist a more general solution which allows to have equal dynamical dimensions.

In this book we analyse a possibility for the massless Thirring model to be renormalisable in the sense of Jackiw's analysis of 1+1 dimensional quantum field theories. Such a possibility can exist only for the more general solutions of the massless Thirring model containing more than one free parameter. Below we show that the dynamical dimensions of Thirring fermion fields can be parametrised by two free parameters, which allow to get equal dynamical dimensions and provide the renormalisability of the massless Thirring model in the sense of Jackiw.

The massless Thirring model is a quantum field theoretic model of fermions with a non–trivial four-fermion interaction in 1+1–dimensional space–time. It is defined by the Lagrangian

$$\mathcal{L}_{\text{Th}}(x) = \bar{\psi}(x)i\gamma^\mu\partial_\mu\psi(x) - \frac{1}{2}g\,\bar{\psi}(x)\gamma^\mu\psi(x)\bar{\psi}(x)\gamma_\mu\psi(x)\,, \tag{1.1}$$

with the massless Dirac fermion field $\psi(x)$ and the dimensionless coupling constant g, which can be both positive or negative. The Lagrangian Eq. (1.1) is invariant under the chiral group $U_V(1) \times U_A(1)$

$$\begin{aligned} V &: \psi(x) \longrightarrow \psi'(x) = e^{i\alpha_V}\psi(x)\,,\\ A &: \psi(x) \longrightarrow \psi'(x) = e^{i\alpha_A\gamma^5}\psi(x)\,, \end{aligned} \tag{1.2}$$

with real constant parameters α_V and α_A. The corresponding Noether currents are given by

$$\begin{aligned} V &: j^\mu(x) = \bar{\psi}(x)\gamma^\mu\psi(x)\,,\\ A &: j_5^\mu(x) = \bar{\psi}(x)\gamma^\mu\gamma^5\psi(x)\,, \end{aligned} \tag{1.3}$$

with $\partial_\mu j^\mu(x) = \partial_\mu j_5^\mu(x) = 0$. Due to Eq. (A.7) they are related as

$$j_5^\mu(x) = -\varepsilon^{\mu\nu} j_\nu(x). \tag{1.4}$$

According to Hagen, the four–fermion interaction in Eq. (1.1) can be linearised by means of the external vector field $A_\mu(x)$. This reduces the massless Thirring model to the massless Schwinger model. Extending Hagen's regularisation procedure we show that the vacuum expectation value of the vector current is proportional to the vector field $A_\mu(x)$ and can be parametrised by two free parameters instead of one as has been pointed out by Hagen within his regularisation procedure. The parametrisation of the vacuum expectation value of the vector current by two free parameters agrees well with the results obtained by Harada et al. [22] (see also Refs. [23, 24]) for the functional fermionic determinant. We denote these parameters as ζ and χ. For $\chi = 1$ the parameter ζ coincides with Hagen's parameter ξ, i.e. $\zeta = \xi$. This means that Hagen's solution as well as Klaiber's solutions are special cases of our more general solution.

1.2. Generating functional of Green functions

The generating functional of correlation functions is defined by

$$Z_{\text{Th}}^{(g)}[J, \bar{J}] = \int \mathcal{D}\psi \mathcal{D}\bar{\psi} \exp\left\{ i \int d^2x \left[\bar{\psi}(x) i\gamma^\mu \partial_\mu \psi(x) - \frac{1}{2} g \, \bar{\psi}(x)\gamma^\mu \psi(x) \bar{\psi}(x) \gamma_\mu \psi(x) \right.\right.$$
$$\left.\left. + \bar{\psi}(x) J(x) + \bar{J}(x) \psi(x) \right] \right\}, \tag{1.5}$$

where $\bar{J}(x)$ and $\bar{J}(x)$ are external sources of the fields $\psi(x)$ and $\bar{\psi}(x)$, respectively. Introducing the external vector source $A_\mu(x)$ of the vector current the generating functional of Green functions can be defined as follows

$$Z_{\text{Th}}^{(g)}[A; J, \bar{J}] = \exp\left\{ \frac{i}{2} g \int d^2x \, \frac{\delta}{\delta A_\mu(x)} \frac{\delta}{\delta A^\mu(x)} \right\} Z_{\text{Th}}^{(0)}[A; J, \bar{J}], \tag{1.6}$$

where we have defined

$$Z_{\text{Th}}^{(0)}[A; J, \bar{J}] = \int \mathcal{D}\psi \mathcal{D}\bar{\psi} \exp\left\{ i \int d^2x \left[\bar{\psi}(x) i\gamma^\mu \partial_\mu \psi(x) + \bar{\psi}(x)\gamma^\mu \psi(x) A_\mu(x) \right.\right.$$
$$\left.\left. + \bar{\psi}(x) J(x) + \bar{J}(x) \psi(x) \right] \right\}. \tag{1.7}$$

This is a generating functional of Green functions in the Schwinger model with massless fermions coupled to an external vector field $A_\mu(x)$. Since it is defined by a Gaussian integral, it can be calculated explicitly. Integration over the fermion fields yields

$$Z_{\text{Th}}^{(0)}[A; J, \bar{J}] = \text{Det}(i\hat{\partial} + \hat{A}) \exp\left\{ i \iint d^2x \, d^2y \, \bar{J}(x) \, S_A(x, y) \, J(y) \right\}, \tag{1.8}$$

where the two–point causal fermion Green function $S_A(x,y)$ is defined by

$$i\gamma^\mu\left(\frac{\partial}{\partial x^\mu} - iA_\mu(x)\right)S_A(x,y) = -\delta^{(2)}(x-y).\tag{1.9}$$

The solution of equation (1.9) is derived in Appendix A.1.1. Here we adduce only the result which reads

$$S_A(x,y) = \frac{1}{2\pi}\frac{\gamma^\mu(x-y)_\mu}{(x-y)^2 - i0}$$
$$\times \exp\left\{-i(g^{\alpha\beta} - \varepsilon^{\alpha\beta}\gamma^5)\int d^2z\,\frac{\partial}{\partial z^\alpha}[\Delta_F(x-z;\mu) - \Delta_F(y-z;\mu)]A_\beta(z)\right\},\tag{1.10}$$

with the antisymmetric tensor $\varepsilon^{\alpha\beta}$, where $\varepsilon^{01} = 1$. Using the relation

$$\frac{\delta}{\delta A_\mu(x)}Z_{\text{Th}}^{(0)}[A;0,0] = i\,\langle 0|j_\mu(x)|0\rangle_{A;g=J=\bar{J}=0},\tag{1.11}$$

and the definition

$$\langle j_\mu(x)\rangle_A = \frac{\langle 0|j_\mu(x)|0\rangle_{A;g=J=\bar{J}=0}}{Z_{\text{Th}}^{(0)}[A;0,0]},\tag{1.12}$$

we follow Hagen for the calculation of the functional determinant $\text{Det}(i\hat{\partial} + \hat{A})$ and calculate the average value of the vector current

$$\langle j_\mu(x)\rangle_A = \frac{1}{i}\frac{\delta}{\delta A^\mu(x)}\ln Z_{\text{Th}}^{(0)}[A;0,0]$$
$$= \frac{1}{i}\frac{\delta}{\delta A^\mu(x)}\ln\text{Det}(i\hat{\partial}+\hat{A}).\tag{1.13}$$

The calculation and regularisation via a spatial point–slitting technique of the averaged value of the vector current is given in Appendices A.1.2 and A.1.3. For the averaged value of the vector current we have found the following expression

$$\langle j_\mu(x)\rangle_A = \int d^2z\,D_{\mu\nu}^{(0)}(x-z)A^\nu(z),\tag{1.14}$$

with

$$D_{\mu\nu}^{(0)}(x-y) = \frac{\zeta}{\pi}g_{\mu\nu}\delta^{(2)}(x-y) - \frac{\chi}{\pi}\frac{\partial}{\partial x^\mu}\frac{\partial}{\partial x^\nu}\Delta_F(x-y;\mu).\tag{1.15}$$

Here ζ and χ are two free parameters, $g_{\mu\nu}$ is the metric tensor and $\Delta_F(x-y;\mu)$ is the causal two–point Green function obeying the equation

$$\Box\Delta_F(x-y;\mu) = \delta^{(2)}(x-y),\tag{1.16}$$

where μ is an infrared cut–off. This Green function is derived in Appendix A.1.1. Since the average of the vector current can be written as

$$\langle j_\mu(x)\rangle_A = \int d^2z\, D^{(0)}_{\mu\nu}(x-z)A^\nu(z)$$
$$= \frac{1}{i}\frac{\delta}{\delta A^\mu(x)}\left[\frac{i}{2}\int d^2z_1 d^2z_2\, A^\sigma(z_1) D^{(0)}_{\sigma\nu}(z_1-z_2)A^\nu(z_2)\right], \quad (1.17)$$

we find the fermionic determinant by comparison of the above relation with Eq. (1.13)

$$\mathrm{Det}(i\hat{\partial}+\hat{A}) = \exp\left\{\frac{i}{2}\int d^2z_1 d^2z_2\, A^\mu(z_1) D^{(0)}_{\mu\nu}(z_1-z_2)A^\nu(z_2)\right\}. \quad (1.18)$$

Finally, the generating functional of fermion fields in the massless Thirring model reads

$$Z^{(g)}_{\mathrm{Th}}[A;J,\bar{J}] = \exp\left\{\frac{i}{2}g\int d^2x\, \frac{\delta}{\delta A_\mu(x)}\frac{\delta}{\delta A^\mu(x)}\right\} Z^{(0)}_{\mathrm{Th}}[A;J,\bar{J}], \quad (1.19)$$

where

$$Z^{(0)}_{\mathrm{Th}}[A;J,\bar{J}] = \exp\left\{\frac{i}{2}\int d^2z_1 d^2z_2\, A^\mu(z_1) D^{(0)}_{\mu\nu}(z_1-z_2)A^\nu(z_2)\right\}$$
$$\times \exp\left\{i\iint d^2x\, d^2y\, \bar{J}(x)\, S_A(x,y)\, J(y)\right\}, \quad (1.20)$$

with $S_A(x,y)$ and $D^{\mu\nu}(z_1-z_2)$ being defined in Eqs. (1.10) and (1.15), respectively. Setting $J(x)=\bar{J}(x)=0$ the generating functional Eq. (1.19) can be calculated explicitly (see Appendix A.1.5). The result reads

$$Z^{(g)}_{\mathrm{Th}}[A;0,0] = \exp\left\{\frac{i}{2}\int d^2z_1 d^2z_2\, A^\mu(z_1) D^{(g)}_{\mu\nu}(z_1-z_2)A^\nu(z_2)\right\}, \quad (1.21)$$

where

$$D^{(g)}_{\mu\nu} = \frac{\zeta}{\pi}\frac{g_{\mu\nu}}{1+\zeta\frac{g}{\pi}}\delta^{(2)}(x-y) - \frac{\chi}{\pi}\frac{1}{\left(1+\zeta\frac{g}{\pi}\right)\left(1+(\zeta-\chi)\frac{g}{\pi}\right)}\frac{\partial}{\partial x^\mu}\frac{\partial}{\partial x^\nu}\Delta_F(x-y;\mu). \quad (1.22)$$

By using the results of this chapter we can proceed to the calculation of the two-point Green function of Thirring fermion fields and two-point correlation function of left-right fermion densities. We investigate these different functions in order to evaluate the dynamical dimensions of the massless Thirring fermion fields and analyse the possibility to make them equal [11].

1.3. Two–point causal Green function $G(x,y)$

The two–point Green function $G(x,y)$ is defined by

$$G(x,y) = i\,\langle 0|\mathrm{T}(\psi(x)\bar{\psi}(y))|0\rangle. \quad (1.23)$$

In terms of the generating functional derived in the last section the Green function reads

$$
\begin{aligned}
G(x,y) &= i\left(\frac{1}{i}\frac{\delta}{\delta \bar{J}(x)}\right)\left(i\frac{\delta}{\delta J(y)}\right)Z^{(g)}_{\text{Th}}[A;J,\bar{J}]\Big|_{A=J=\bar{J}=0} \\
&= \frac{1}{i}\frac{\delta}{\delta J(y)}\frac{\delta}{\delta \bar{J}(x)}Z^{(g)}_{\text{Th}}[A;J,\bar{J}]\Big|_{A=J=\bar{J}=0} \\
&= \exp\left\{\frac{i}{2}g\int d^2z\, \frac{\delta}{\delta A_\mu(z)}\frac{\delta}{\delta A^\mu(z)}\right\} \\
&\quad \times \exp\left\{\frac{i}{2}\int d^2z_1 d^2z_2\, A^\mu(z_1)D^{(0)}_{\mu\nu}(z_1-z_2)A^\nu(z_2)\right\}S_A(x,y)\Big|_{A=0}.
\end{aligned}
\tag{1.24}
$$

The calculation is performed in Appendix A.2.1 and the result is

$$
\begin{aligned}
G(x,y) &= \frac{1}{2\pi}\frac{\gamma^\mu(x-y)_\mu}{(x-y)^2-i0} \\
&\quad \times \exp\left\{\frac{ig^2}{\pi}\frac{\chi}{\left(1+(\zeta-\chi)\frac{g}{\pi}\right)\left(1+\zeta\frac{g}{\pi}\right)}\Big(\Delta_F(0;\mu)-\Delta_F(x-y;\mu)\Big)\right\}.
\end{aligned}
\tag{1.25}
$$

The Green function $\Delta_F(0;\mu)$ is evaluated in Appendix A.2.2 and reads

$$
\Delta_F(0;\mu) = -\frac{1}{4\pi i}\ln\left(\frac{\Lambda^2}{\mu^2}\right).
\tag{1.26}
$$

Introducing the dynamical dimension of the Thirring fermion field defined by [11]

$$
d_G = \frac{g^2}{4\pi^2}\frac{\chi}{\left(1+\zeta\frac{g}{\pi}\right)\left(1+(\zeta-\chi)\frac{g}{\pi}\right)},
\tag{1.27}
$$

the two–point Green function is found to be

$$
\begin{aligned}
G(x,y) &= \frac{1}{2\pi}\frac{\gamma^\mu(x-y)_\mu}{(x-y)^2-i0}\exp\left\{4\pi i\, d_G\Big(\Delta_F(0;\mu)-\Delta_F(x-y;\mu)\Big)\right\} \\
&= -\frac{\Lambda^2}{2\pi}\frac{\gamma^\mu(x-y)_\mu}{-\Lambda^2(x-y)^2+i0}\Big(-\Lambda^2(x-y)^2+i0\Big)^{-d_G} \\
&= \Lambda\, G(d_G;\Lambda x,\Lambda y).
\end{aligned}
\tag{1.28}
$$

1.4. Two–point correlation function $C(x,y)$

The two–point correlation function $C(x,y)$ of the left–right fermion densities is defined by

$$
C(x,y) = \left\langle 0\left|\text{T}\left(\bar{\psi}(x)\left(\frac{1-\gamma^5}{2}\right)\psi(x)\,\bar{\psi}(y)\left(\frac{1+\gamma^5}{2}\right)\psi(y)\right)\right|0\right\rangle.
\tag{1.29}
$$

Using the generating functional we find

$$
\begin{aligned}
C(x,y) &= \mathrm{Tr}\left\{\left(i\frac{\delta}{\delta J(x)}\right)\left(\frac{1-\gamma^5}{2}\right)\left(\frac{1}{i}\frac{\delta}{\delta \bar{J}(x)}\right)\right\} \\
&\quad \times \mathrm{Tr}\left\{\left(i\frac{\delta}{\delta J(y)}\right)\left(\frac{1+\gamma^5}{2}\right)\left(\frac{1}{i}\frac{\delta}{\delta \bar{J}(y)}\right)\right\}Z^{(g)}_{\mathrm{Th}}[A;J,\bar{J}]\bigg|_{A=J=\bar{J}=0} \\
&= \exp\left\{\frac{i}{2}g\int d^2z\,\frac{\delta}{\delta A_\mu(z)}\frac{\delta}{\delta A^\mu(z)}\right\}\exp\left\{\frac{i}{2}\int d^2z_1 d^2z_2\,A^\mu(z_1)D^{(0)}_{\mu\nu}(z_1-z_2)A^\nu(z_2)\right\} \\
&\quad \times \mathrm{Tr}\left\{\left(\frac{1-\gamma^5}{2}\right)S_A(x,y)\left(\frac{1+\gamma^5}{2}\right)S_A(y,x)\right\}\bigg|_{A=0}. \quad (1.30)
\end{aligned}
$$

The calculations are performed in Appendix A.3 yielding

$$
C(x,y) = -\frac{1}{4\pi^2}\frac{1}{(x-y)^2-i0}\exp\left\{-\frac{4ig}{1+\zeta\frac{g}{\pi}}\Big(\Delta_F(0;\mu)-\Delta_F(x-y;\mu)\Big)\right\}. \quad (1.31)
$$

Using Eq. (1.26) again and introducing the dynamical dimension d_C

$$
d_C = -\frac{g}{2\pi}\frac{1}{1+\zeta\frac{g}{\pi}}, \quad (1.32)
$$

the correlation function reads

$$
\begin{aligned}
C(x,y) &= -\frac{1}{4\pi^2}\frac{1}{(x-y)^2-i0}\exp\left\{8\pi d_C\Big(i\Delta_F(0;\mu)-i\Delta_F(x-y;\mu)\Big)\right\} \\
&= \frac{\Lambda^2}{4\pi^2}\frac{1}{-\Lambda^2(x-y)^2+i0}\Big(-\Lambda^2(x-y)^2+i0\Big)^{-2d_C} \\
&= \Lambda^2\,C(d_C;\Lambda x,\Lambda y). \quad (1.33)
\end{aligned}
$$

Equating d_C to d_G, $d_C=d_G$, which is necessary for the renormalisation of the massless Thirring model we get the constraint on the parameters

$$
\chi = \frac{2\pi}{g}\left(1+\zeta\frac{g}{\pi}\right). \quad (1.34)
$$

Below we show that positive definiteness of the norms of the wave functions of the states related to the components of the vector current does not violate such a relation. We also corroborate the necessity of the condition $d_C=d_G$ within the standard renormalisation scheme.

1.5. Non–perturbative renormalisation of the massless Thirring model

Non-perturbative renormalisability of the massless Thirring model is understood as the possibility to remove all ultraviolet and infrared divergences by renormalisation of the fermion field $\psi(x)$ and the coupling constant g. As usual in the renormalisation procedure [20] the

Lagrangian is rewritten in terms of bare quantities

$$\mathcal{L}_{\text{Th}}(x) = \bar{\psi}_0(x)i\gamma^\mu\partial_\mu\psi_0(x) - \frac{1}{2}g_0\,\bar{\psi}_0(x)\gamma^\mu\psi_0(x)\bar{\psi}_0(x)\gamma_\mu\psi_0(x)\,, \tag{1.35}$$

with the bare field operators $\psi_0(x)$, $\bar{\psi}_0(x)$ and the bare coupling constant g_0. Expressing the Lagrangian in renormalised quantities

$$\begin{aligned}\mathcal{L}_{\text{Th}}(x) &= \bar{\psi}(x)i\gamma^\mu\partial_\mu\psi(x) - \frac{1}{2}g\,\bar{\psi}(x)\gamma^\mu\psi(x)\bar{\psi}(x)\gamma_\mu\psi(x) \\ &\quad + (Z_2-1)\,\bar{\psi}(x)i\gamma^\mu\partial_\mu\psi(x) - \frac{1}{2}g\,(Z_1-1)\,\bar{\psi}(x)\gamma^\mu\psi(x)\bar{\psi}(x)\gamma_\mu\psi(x) \\ &= Z_2\,\bar{\psi}(x)i\gamma^\mu\partial_\mu\psi(x) - \frac{1}{2}g\,Z_1\,\bar{\psi}(x)\gamma^\mu\psi(x)\bar{\psi}(x)\gamma_\mu\psi(x)\,,\end{aligned} \tag{1.36}$$

where Z_1 and Z_2 are renormalisation constants, we find for the relation between renormalised and bare quantities

$$\psi_0(x) = Z_2^{1/2}\,\psi(x)\,, \tag{1.37}$$

$$g_0 = Z_1 Z_2^{-2}\,g\,. \tag{1.38}$$

For the correlation functions of massless Thirring fermions the renormalisability of the massless Thirring model is the possibility to replace the infrared cut–off μ and the ultraviolet cut–off Λ by a finite scale M by means of the renormalisation constants Z_1 and Z_2.

The bare n-point correlation functions with cut–off Λ can be expressed by the correlation functions at a finite scale M

$$G_0(x_1,\ldots x_n,y_1,\ldots,y_n;\Lambda) = \left(\frac{\Lambda}{M}\right)^{-2nd_G}G_0(x_1,\ldots x_n,y_1,\ldots,y_n;M)\,,$$

$$C_0(x_1,\ldots x_n,y_1,\ldots,y_n;\Lambda) = \left(\frac{\Lambda}{M}\right)^{-4nd_C}C_0(x_1,\ldots x_n,y_1,\ldots,y_n;M)\,. \tag{1.39}$$

Due to wave function renormalisation Eq. (1.37) the correlation functions renormalised at the scale M read

$$\begin{aligned}G(x_1,\ldots x_n,y_1,\ldots,y_n) &= Z_2^{-n}\,G_0(x_1,\ldots x_n,y_1,\ldots,y_n;\Lambda) \\ &= Z_2^{-n}\left(\frac{\Lambda}{M}\right)^{-2nd_G}G_0(x_1,\ldots x_n,y_1,\ldots,y_n;M)\,, \\ C(x_1,\ldots x_n,y_1,\ldots,y_n) &= Z_2^{-2n}\,C_0(x_1,\ldots x_n,y_1,\ldots,y_n;\Lambda) \\ &= Z_2^{-2n}\left(\frac{\Lambda}{M}\right)^{-4nd_C}C_0(x_1,\ldots x_n,y_1,\ldots,y_n;M)\,.\end{aligned} \tag{1.40}$$

Renormalisation at the scale M demands

$$Z_2 = \left(\frac{\Lambda}{M}\right)^{-2d_G}\,,$$

$$Z_2 = \left(\frac{\Lambda}{M}\right)^{-2d_C}. \tag{1.41}$$

One can see that renormalisation is possible only for equal dynamical dimensions, i.e. $d_G = d_C$. Thus we find that the coupling constant g_0 is not affected by quantum corrections making its renormalisation trivial. So we are free to set $Z_1 = Z_2^2$ yielding $g_0 = g$. Hence we find that the massless Thirring model is non–perturbative renormalisable.

1.6. The Schwinger term

In this section we evaluate the Schwinger term Ref. [21] for the free theory ($g = 0$). Since

$$[j^0(x), j^1(y)]_{x^0=y^0} = j^0(x)j^1(y)|_{x^0=y^0} - j^1(y)j^0(x)|_{x^0=y^0}$$
$$= \lim_{x^0 \to y^0+} \langle 0|T(j^0(x)j^1(y))|0\rangle - \lim_{x^0 \to y^0-} \langle 0|T(j^0(x)j^1(y))|0\rangle, \tag{1.42}$$

we have to find the current correlation function. The correlation function is obtained by the relation

$$\frac{1}{Z_{\text{Th}}^{(0)}[A;0,0]} \frac{1}{i^2} \frac{\delta}{\delta A_\mu(x)} \frac{\delta}{\delta A_\nu(y)} Z_{\text{Th}}^{(0)}[A;0,0]\bigg|_{A=0} = \langle 0|T(j^\mu(x)j^\nu(y))|0\rangle - i\langle 0|\frac{\delta j^\mu(x)}{\delta A_\nu(y)}|0\rangle, \tag{1.43}$$

respectively

$$\langle 0|T(j^0(x)j^1(y))|0\rangle = \frac{1}{Z_{\text{Th}}^{(0)}[A;0,0]} \frac{1}{i^2} \frac{\delta}{\delta A_0(x)} \frac{\delta}{\delta A_1(y)} Z_{\text{Th}}^{(0)}[A;0,0]\bigg|_{A=0} + i\langle 0|\frac{\delta j^0(x)}{\delta A_1(y)}|0\rangle. \tag{1.44}$$

The partition function Eq. (1.20) reads in our case

$$Z_{\text{Th}}^{(0)}[A;0,0] = \exp\left\{\frac{i}{2}\int d^2z_1 d^2z_2\, A^\mu(z_1) D_{\mu\nu}^{(0)}(z_1-z_2) A^\nu(z_2)\right\}, \tag{1.45}$$

so we obtain

$$\frac{1}{Z_{\text{Th}}^{(0)}[A;0,0]} \frac{1}{i^2} \frac{\delta}{\delta A^0(x)} \frac{\delta}{\delta A^1(y)} Z_{\text{Th}}^{(0)}[A;0,0]\bigg|_{A=0} = -iD_{01}^{(0)}(x-y)$$
$$= i\frac{\chi}{\pi} \frac{\partial}{\partial x^0} \frac{\partial}{\partial x^1} \Delta_F(x-y;\mu), \tag{1.46}$$

where we have used Eq. (1.15) in the last line. The derivative of the current reads (see Eq. (A.55))

$$i\langle 0|\frac{\delta j^0(x)}{\delta A_1(y)}|0\rangle = -i\frac{\chi-1}{\pi} \frac{\partial}{\partial x_0} \frac{\partial}{\partial x_1} \Delta_F(x-y;\mu), \tag{1.47}$$

and the current relation is given by

$$\langle 0|T(j^0(x)j^1(y))|0\rangle = \frac{i}{\pi}\frac{\partial}{\partial x_0}\frac{\partial}{\partial x_1}\Delta_F(x-y;\mu). \tag{1.48}$$

For the Schwinger term we find

$$[j^0(x), j^1(y)]_{x^0=y^0} = \frac{i}{\pi}\left(\lim_{x^0\to y^0+} - \lim_{x^0\to y^0-}\right)\frac{\partial}{\partial x_0}\frac{\partial}{\partial x_1}\Delta_F(x-y;\mu)$$

$$= -\frac{i}{\pi}\frac{\partial}{\partial x^1}\delta(x^1-y^1), \tag{1.49}$$

where the difference of the limits is performed in Appendix A.4 and the index 1 was raised in the derivative yielding an additional minus.

1.7. The norm of the wave functions of the states related to the components of the vector current

1.7.1. The free theory

The current correlation function is obtained by the relation

$$\langle 0|T(j^\mu(x)j^\nu(y))|0\rangle = \frac{1}{Z_{\text{Th}}^{(0)}[A;0,0]}\frac{1}{i^2}\frac{\delta}{\delta A_\mu(x)}\frac{\delta}{\delta A_\nu(y)}Z_{\text{Th}}^{(0)}[A;0,0]\bigg|_{A=0} + i\langle 0|\frac{\delta j^\mu(x)}{\delta A_\nu(y)}|0\rangle$$

$$= \frac{i}{\pi}\left(-g^{\mu 0}g^{\nu 0}\delta^{(2)}(x-y) + \frac{\partial}{\partial x_\mu}\frac{\partial}{\partial x_\nu}\Delta_F(x-y;\mu)\right), \tag{1.50}$$

where we have used Eq. (A.55). This result is corroborated in Appendix A.5.1 for $x \neq y$. Using the Wightman function

$$D^{(\pm)}(x-y) = \int\frac{d^2k}{(2\pi)^2}2\pi\theta(k^0)\delta(k^2)\,e^{\mp ik\cdot(x-y)}$$

$$= \frac{1}{4\pi}\int\frac{dk^1}{\sqrt{k_1^2+\mu^2}}e^{\mp i\sqrt{k_1^2+\mu^2}(x^0-y^0)\pm ik^1(x^1-y^1)}, \tag{1.51}$$

it is shown in Appendix A.5.2 that the current operator product reads

$$\langle 0|j^\mu(x)j^\nu(y)|0\rangle = -\frac{1}{\pi}\frac{\partial}{\partial x_\mu}\frac{\partial}{\partial x_\nu}D^{(+)}(x-y). \tag{1.52}$$

The vacuum expectation values are

$$\langle 0|j^0(x)j^0(y)|0\rangle = -\frac{1}{\pi}\left(\frac{\partial}{\partial x^1}\right)^2 D^{(+)}(x-y),$$

$$\langle 0|j^1(x)j^1(y)|0\rangle = -\frac{1}{\pi}\left(\frac{\partial}{\partial x^1}\right)^2 D^{(+)}(x-y), \tag{1.53}$$

where we have used $\Box D^{(+)}(x-y) = 0$ in the first line. According to Wightman and Streater [49] and Coleman [50], we can define the wave functions of the states

$$|h; j^0\rangle = \int d^2x\, h(x)\, j^0(x)|0\rangle,$$
$$|h; j^1\rangle = \int d^2x\, h(x)\, j^1(x)|0\rangle, \qquad (1.54)$$

with $h(x)$ being a test function from the Schwartz class $h(x) \in \mathcal{S}(\mathbb{R}^2)$ [49].

The norms of the states (1.54) read [49, 50]

$$\langle j^0; h|h; j^0\rangle = \int d^2x \int d^2y\, h^*(x)\, \langle 0|j^0(x)j^0(y)|0\rangle\, h(y)$$
$$= \frac{1}{\pi} \int \frac{d^2k}{(2\pi)^2}\, 2\pi\, (k^0)^2 \theta(k^0) \delta(k^2)\, |\tilde{h}(k)|^2,$$
$$\langle j^1; h|h; j^1\rangle = \int d^2x \int d^2y\, h^*(x)\, \langle 0|j^1(x)j^1(y)|0\rangle\, h(y)$$
$$= \frac{1}{\pi} \int \frac{d^2k}{(2\pi)^2}\, 2\pi\, (k^0)^2 \theta(k^0) \delta(k^2)\, |\tilde{h}(k)|^2, \qquad (1.55)$$

where $\tilde{h}(k)$ is the Fourier transform of the test function $h(x)$. One can see that the norms are indeed positive.

1.7.2. The interacting theory

In this section we define the constraints on the parameters ζ and χ caused by the positive definiteness of the norms of the wave functions of the states related to the components of the vector current. For this aim we need the two-point correlation function $\langle 0|T(j^\mu(x)j^\nu(y))|0\rangle$. According to Johnson [5], this correlation function takes the form (compare to the free theory)

$$i\langle 0|T(j^\mu(x)j^\nu(y))|0\rangle = -\frac{\chi}{\pi} \frac{1}{\left(1+\zeta\frac{g}{\pi}\right)\left(1+(\zeta-\chi)\frac{g}{\pi}\right)} \frac{\partial}{\partial x_\mu}\frac{\partial}{\partial x_\nu} \Delta_F(x-y;\mu)$$
$$+ \frac{\chi}{\pi} \frac{1}{\left(1+\zeta\frac{g}{\pi}\right)\left(1+(\zeta-\chi)\frac{g}{\pi}\right)} g^{\mu 0} g^{\nu 0} \delta^{(2)}(x-y). \qquad (1.56)$$

This yields the following expressions for the vacuum expectation values $\langle 0|j^0(x)j^0(y)|0\rangle$ and $\langle 0|j^1(x)j^1(y)|0\rangle$

$$\langle 0|j^0(x)j^0(y)|0\rangle = -\frac{\chi}{\pi} \frac{1}{\left(1+\zeta\frac{g}{\pi}\right)\left(1+(\zeta-\chi)\frac{g}{\pi}\right)} \left(\frac{\partial}{\partial x^1}\right)^2 D^{(+)}(x-y),$$
$$\langle 0|j^1(x)j^1(y)|0\rangle = -\frac{\chi}{\pi} \frac{1}{\left(1+\zeta\frac{g}{\pi}\right)\left(1+(\zeta-\chi)\frac{g}{\pi}\right)} \left(\frac{\partial}{\partial x^1}\right)^2 D^{(+)}(x-y), \qquad (1.57)$$

with $D^{(\pm)}(x-y)$ being the Wightman functions

$$D^{(\pm)}(x-y) = \int \frac{d^2k}{(2\pi)^2}\, 2\pi \theta(k^0)\delta(k^2)\, e^{\mp ik\cdot(x-y)}, \qquad (1.58)$$

obeying the relation

$$\Delta_F(x-y;\mu) = i\theta(x^0-y^0)\, D^{(+)}(x-y) + i\theta(y^0-x^0)\, D^{(-)}(x-y). \qquad (1.59)$$

Following again Wightman and Streater [49] and Coleman [50], we take the wave functions of the states related to the components of the vector current as

$$|h;j^0\rangle = \int d^2x\, h(x)\, j^0(x)|0\rangle,$$
$$|h;j^1\rangle = \int d^2x\, h(x)\, j^1(x)|0\rangle, \qquad (1.60)$$

with $h(x)$ being a test function from the Schwartz class $h(x) \in \mathcal{S}(\mathbb{R}^2)$ [49].

The norms of the states (1.60) read [49, 50]

$$\langle j^0;h|h;j^0\rangle = \int d^2x \int d^2y\, h^*(x)\, \langle 0|j^0(x)j^0(y)|0\rangle\, h(y)$$
$$= \frac{\chi}{\pi}\, \frac{1}{\left(1+\zeta\frac{g}{\pi}\right)\left(1+(\zeta-\chi)\frac{g}{\pi}\right)} \int \frac{d^2k}{(2\pi)^2}\, 2\pi\, (k^0)^2 \theta(k^0)\delta(k^2)\, |\tilde{h}(k)|^2,$$
$$\langle j^1;h|h;j^1\rangle = \int d^2x \int d^2y\, h^*(x)\, \langle 0|j^1(x)j^1(y)|0\rangle\, h(y)$$
$$= \frac{\chi}{\pi}\, \frac{1}{\left(1+\zeta\frac{g}{\pi}\right)\left(1+(\zeta-\chi)\frac{g}{\pi}\right)} \int \frac{d^2k}{(2\pi)^2}\, 2\pi\, (k^0)^2 \theta(k^0)\delta(k^2)\, |\tilde{h}(k)|^2, \qquad (1.61)$$

with $\tilde{h}(k)$ being the Fourier transform of the test function $h(x)$. The positive definiteness of the norms yields the constraint

$$\chi\left(1+\zeta\frac{g}{\pi}\right)\left(1+(\zeta-\chi)\frac{g}{\pi}\right) > 0. \qquad (1.62)$$

This assumes that $\chi \neq 0$. Using the constraint on the renormalisability of the massless Thirring model (1.34), the inequality (1.62) reads

$$-g\left(1+\zeta\frac{g}{\pi}\right) > 0. \qquad (1.63)$$

This inequality is fulfilled for

$$1+\zeta\frac{g}{\pi} < 0 \quad \text{if} \quad g > 0, \qquad (1.64)$$
$$1+\zeta\frac{g}{\pi} > 0 \quad \text{if} \quad g < 0. \qquad (1.65)$$

The inequality (1.62) yields the following interesting consequences. According to Coleman [30], the coupling constant β^2 of the sine–Gordon model is related to the coupling constant

g of the Thirring model as

$$\begin{aligned}\frac{\beta^2}{8\pi} &= \frac{1}{2} + d_C(g) \\ &= \frac{1}{2}\left(1 - \frac{g}{\pi}\frac{1}{1+\zeta\frac{g}{\pi}}\right).\end{aligned} \quad (1.66)$$

Therefore, for the constraint (1.64) the coupling constant β^2 is of order $\beta^2 \sim 8\pi$. The renormalisability of the sine–Gordon model at this region of the coupling constant is the topic of the second chapter of this book (see also Ref. [31]).

1.8. Conclusion

We have found a more general solution to the massless Thirring model, which is parametrised by two parameters. These parameters do not appear explicitly in the Lagrangian. Our solutions incorporate those obtained by Johnson, Sommerfield, Hagen, Klaiber and within the path–integral approach [5]–[7], [9], [13]–[19].

As has been pointed out by Jackiw [11], a key–problem of 1+1–dimensional quantum field theories is the inequality of dynamical dimensions of fermion fields obtained from different correlation functions. We have calculated the causal two–point Green function and the two–point correlation function of left–right fermion densities and have shown that their dynamical dimensions d_G and d_C can be made equal. This gives a constraint relating the two parameters. Since all dynamical dimensions of all correlation functions can be made equal the massless Thirring model is renormalisable in the sense that the dependence of correlation functions of Thirring fermion fields on the ultraviolet cut–off can be removed by renormalisation of the wave function of Thirring fermion fields only. The necessity to have equal dimensions for renormalisability of the massless Thirring model was shown within the standard renormalisation procedure.

Finally, the Schwinger term and the norm of the current correlation function of the free and interacting theory were calculated. The norms of the wave functions of the states related to the vector current impose some constraints on two parameters of this approach but do not contradict to the possibility of renormalising the massless Thirring model.

2. Sine–Gordon Model

2.1. Introduction to the "classical" sine–Gordon model

The sine–Gordon model is a system in 1+1 space–time dimensions which yields solitons among its solutions. The definition of a soliton is mathematically precisely defined (see e.g. Ref. [51]). Loosely speaking, one calls solutions of a system which are non-dissipative lumps of energy extended to a finite region in space and which sustain their shape and velocity *solitary waves* (the name is due to John Scott-Russell who was the first to document the observation of a solitary wave in 1834). If solitary waves sustain their shape and velocity asymptotically even after collisions we speak of *solitons* (Zabusky and Kruskal [52] coined the name soliton for solitary waves with particle properties). The sine–Gordon model has a long history and is well known in pure mathematics and several areas in physics. Its name is a pun on the Klein-Gordon equation and was introduced according to S. Coleman by D. Finkelstein and J. Rubinstein. The latter refers to M. Kruskal as the first author of this name.

An infinite set of massive pendula connected by springs in a homogeneous gravitational field is an example of a classical system which in the continuum limit is described by the sine–Gordon model. Here the sine–Gordon field corresponds to the angle of rotation of all the pendula. Around 1960 the equation entered particle physics due to the work of Perring and Skyrme [53, 54], who considered the sine–Gordon model as a model for baryons. They examined scattering and confirmed the particle-like stability of solitons (see Eqs. (2.26), (2.27)). Furthermore, they defined the topological charge (see Eq. (2.14)). Frenkel and Kontrova [55] were the first who applied the sine–Gordon model to the propagation of crystal dislocations. There, the displacement of atoms connected by linear springs may propagate as a soliton in the periodic crystal field. Seeger, Donth and Kochendorfer [56] discovered the soliton-soliton and soliton-antisoliton solution investigating the propagation of crystal dislocations. McCall and Hahn [57] found that coherent light propagating in 2-level atoms obeys the sine–Gordon equation if the spectral widths are neglected (perfect resonance). The observed soliton behaviour is called self-induced transparency. Furthermore, the sine–Gordon model appears also in phenomena like splay waves in membranes, the propagation of magnetic fluxes in the Josephson junction, Bloch wall motion in magnetic crystals, the edge effect of the Quantum Hall effect, in one-dimensional organic conductors, one-dimensional ferromagnets and liquid crystals. Furthermore, the sine–Gordon model is equivalent to the massive Thirring model [30, 58], while the Schwinger model of two-dimensional QED bosonises to the sine–Gordon model with a mass term added (see Ref. [59, 60]). For details to these applications see Scott et al. [61] and Barone et al. [62].

Besides its wide range of applications maybe the most interesting property of the sine–

Gordon model is its exact solvability which makes it an ideal system to study non-perturbative aspects of quantum field theory, to which the main part of this chapter is devoted. It seems that the sine–Gordon model is the complete essence of quantum physics in 1+1 dimensional flat space–time.

The sine–Gordon model is defined by the Lagrangian

$$\mathcal{L}(x) = \frac{1}{2}\partial_\mu\varphi(x)\partial^\mu\varphi(x) + \frac{\alpha}{\beta^2}\left(\cos\beta\varphi(x) - 1\right),\tag{2.1}$$

yielding the equation of motion, which is called the sine–Gordon equation

$$\Box\varphi(x) + \frac{\alpha}{\beta}\sin\beta\varphi(x) = 0\,.\tag{2.2}$$

Expansion of the interaction part

$$\mathcal{L}(x) = \frac{1}{2}\partial_\mu\varphi(x)\partial^\mu\varphi(x) - \frac{\alpha}{2}\varphi(x)^2 + \frac{\alpha\beta^2}{4!}\varphi(x)^4 + \mathcal{O}(\alpha\beta^4)\,,\tag{2.3}$$

allows to interpret the constants α and β. While the former has the meaning of a mass squared, $\alpha\beta^2$ corresponds to a coupling constant of a quartic interaction. For $\beta = 0$ the Lagrangian reduces to the free Klein-Gordon Lagrangian. After the replacement $\beta\varphi \to \varphi$ we find for the Lagrangian

$$\mathcal{L}(x) = \frac{1}{\beta^2}\left(\frac{1}{2}\partial_\mu\varphi(x)\partial^\mu\varphi(x) + \alpha\left(\cos\varphi(x) - 1\right)\right),\tag{2.4}$$

which reveals an additional meaning of β^2. Since the relevant object in quantum physics is the phase of the path integral, i.e. $i\int d^2x\,\mathcal{L}(x)/\hbar$, we find that changing β^2 in all physical results has exactly the same effect as changing $\hbar$. Therefore, β^2 has the meaning of $\hbar$, which agrees well with the fact that β^2 is not affected by quantum corrections, i.e. is not a renormalised coupling constant.

Using the canonical field momentum

$$\pi = \frac{\partial\mathcal{L}}{\partial\partial_0\varphi} = \dot\varphi\,,\tag{2.5}$$

we find for the energy-momentum tensor

$$\begin{aligned}T^{\mu\nu} &= \frac{\partial\mathcal{L}}{\partial\partial_\mu\varphi}\partial^\nu\varphi - g^{\mu\nu}\mathcal{L}\\ &= \partial^\mu\varphi\partial^\nu\varphi - g^{\mu\nu}\partial^\sigma\varphi\partial_\sigma\varphi - g^{\mu\nu}\frac{\alpha}{\beta^2}\left(\cos\beta\varphi(x) - 1\right),\end{aligned}\tag{2.6}$$

and its components

$$T^{00} = \frac{\dot\varphi^2 + \varphi'^2}{2} - \frac{\alpha}{\beta^2}\left(\cos\beta\varphi(x) - 1\right),\tag{2.7}$$

$$T^{01} = \dot\varphi\varphi' = T^{10}\,,\tag{2.8}$$

2.1. Introduction to the "classical" sine–Gordon model

$$T^{11} = \frac{\dot{\varphi}^2 + \varphi'^2}{2} + \frac{\alpha}{\beta^2}\left(\cos\beta\varphi(x) - 1\right). \tag{2.9}$$

The discrete symmetries of the sine–Gordon system are

$$\varphi(x) \longrightarrow -\varphi(x), \tag{2.10}$$

$$\varphi(x) \longrightarrow \varphi(x) + \frac{2\pi n}{\beta} \qquad n \in \mathbb{N}. \tag{2.11}$$

Therefore, the addition of $2\pi n/\beta$ to a given solution or the change of its global sign yields again a solution of the sine–Gordon model. At the minima of the potential

$$U = \frac{\alpha}{\beta^2}\left(1 - \cos\beta\varphi(x)\right), \tag{2.12}$$

at the field values

$$\varphi(x) = \frac{2\pi n}{\beta} \qquad n \in \mathbb{N}, \tag{2.13}$$

the energy $\int d^3x\, T^{00}$ vanishes. All field configurations with finite energy approach one of the values Eq. (2.13) for $x \to \pm\infty$. Denoting n_L the appropriate integer for $x \to -\infty$ and n_R the corresponding one for $x \to +\infty$ we find that for $n_L = n_R = n$ the lowest energy configuration is given by $\varphi(x) = 2\pi n/\beta$, which has vanishing energy. For $n_L \neq n_R$ the lowest energy configuration is a topologically non-trivial field which obviously is independent of the values n_L and n_R themselves but depends on their difference which is called the *topological charge*

$$Q = n_R - n_L = \frac{\beta}{2\pi}\int_{-\infty}^{+\infty} dx\, \frac{\partial\varphi(x)}{\partial x}. \tag{2.14}$$

Obviously, this charge is constant with respect to time and should therefore correspond to a conserved 2–current $\partial_\mu j^\mu = 0$. Since

$$Q = \int_{-\infty}^{+\infty} dx\, j^0(x), \tag{2.15}$$

with

$$j^0(x) = \frac{\beta}{2\pi}\frac{\partial\varphi(x)}{\partial x} = \frac{\beta}{2\pi}\varepsilon^{01}\partial_1\varphi(x), \tag{2.16}$$

with the epsilon tensor in two dimensions $\varepsilon^{\mu\nu} = -\varepsilon^{\nu\mu}$ and $\varepsilon^{01} = 1$ we find for the *topological current*

$$j^\mu(x) = \frac{\beta}{2\pi}\varepsilon^{\mu\nu}\partial_\nu\varphi(x),$$

$$\partial_\mu j^\mu(x) = 0. \tag{2.17}$$

The conservation of the current is due to topology and not a consequence of the equations of motion, i.e. it is not a Noether current. For a single scalar field in one space dimension (non-trivial) static solutions must have $Q = \pm 1$, which is obvious in the picture of the coupled pendula in a gravitational field stated above. If $|Q| > 1$ the pendula separate in $n > 1$, $n \in \mathbb{N}$ lumps of energy which interact yielding a non-static solution.

The derivation of the $Q = \pm 1$ solution is done as follows. The equation of motion

$$\Box \varphi(x) + U'(\varphi) = 0\,, \tag{2.18}$$

for a static solution becomes

$$\varphi''(x) = U'(\varphi)\,. \tag{2.19}$$

This can be integrated yielding

$$\varphi'^2(x) = 2U(\varphi)\,, \tag{2.20}$$

$$\varphi'(x) = \pm\sqrt{2U(\varphi)}\,, \tag{2.21}$$

and

$$\int_{\varphi(x_0)}^{\varphi(x)} \frac{d\varphi}{\sqrt{2U(\varphi)}} = \pm \int_{x_0}^{x} dx\,. \tag{2.22}$$

Introducing $\theta = \beta\varphi/2$ and using the relation $1 - \cos 2\theta = 2\sin^2\theta$ we find

$$\frac{1}{\sqrt{\alpha}} \int_{\theta(x_0)}^{\theta(x)} \frac{d\theta}{\sin\theta} = \pm \int_{x_0}^{x} dx\,, \tag{2.23}$$

which becomes after integration

$$\frac{1}{\sqrt{\alpha}} \ln\left|\tan\frac{\theta(x)}{2}\right| = \pm(x - x_0)\,, \tag{2.24}$$

where we have made the replacement $x_0 \mp \frac{1}{\sqrt{\alpha}} \ln\left|\tan\frac{\theta(x_0)}{2}\right| \to x_0$. After inversion we find for the solution with $Q = \pm 1$

$$\varphi(x) = \frac{4}{\beta} \arctan\exp\left(\pm\sqrt{\alpha}(x - x_0)\right)\,, \tag{2.25}$$

which is called the *soliton* or *kink* (+) and *antisoliton* or *antikink* (-) in the literature. Due to the Lorentz symmetry of the sine–Gordon model a soliton moving with velocity v is obtained by a Lorentz boost

$$\varphi_S(x,t) = \frac{4}{\beta} \arctan\exp\left(\sqrt{\alpha}\gamma(x - x_0 - vt)\right)\,, \tag{2.26}$$

with the relativistic gamma-factor $\gamma = 1/\sqrt{1-v^2}$ (we have set c=1). The soliton at rest is

2.1. Introduction to the "classical" sine–Gordon model

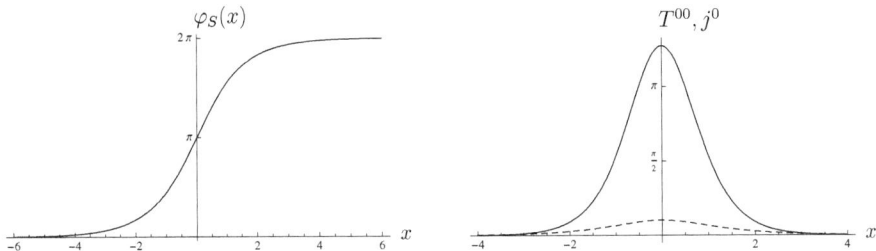

Figure 2.1.: On the *left* the soliton solution $\varphi_S(x,t)$ with the parameters $x_0 = 0$, $\alpha = \beta = 1$ is plotted while on the *right* the corresponding energy density T^{00} is the solid line and the current density j^0 the dashed one.

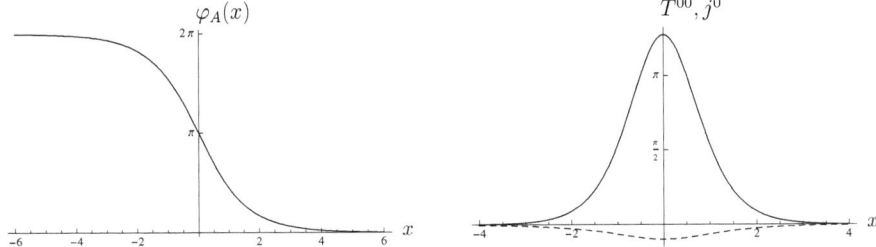

Figure 2.2.: On the *left* the antisoliton solution $\varphi_A(x,t)$ with the parameters $x_0 = 0$, $\alpha = \beta = 1$ is plotted while on the *right* the corresponding energy density T^{00} is plotted as solid line and the current density j^0 as the dashed one.

depicted in Fig. 2.1. Using the relation $\arctan \exp(-x) = -\arctan \exp(x) + \pi/2$ and Eq. (2.11) we find for the antisoliton (see Fig. 2.2)

$$\varphi_A(x,t) = -\frac{4}{\beta} \arctan \exp\left(\sqrt{\alpha}\gamma(x - x_0 - vt)\right) = -\varphi_S(x,t). \tag{2.27}$$

The energy of a soliton is obtained as follows. Using Eq. (2.20) we find for a soliton at rest

$$\begin{aligned} T^{00} &= \varphi'^2 \\ &= \left(\frac{4\sqrt{\alpha}}{\beta} \frac{e^{\sqrt{\alpha}x}}{1 + e^{2\sqrt{\alpha}x}}\right)^2 \\ &= \frac{16\alpha}{\beta^2} \frac{e^{2\sqrt{\alpha}x}}{(1 + e^{2\sqrt{\alpha}x})^2}, \end{aligned} \tag{2.28}$$

where we have set $x_0 = 0$. The spatial integration can be performed easily and we find

$$\begin{aligned} E &= \int dx \, T^{00} \\ &= \int dx \, \frac{16\alpha}{\beta^2} \frac{e^{2\sqrt{\alpha}x}}{(1 + e^{2\sqrt{\alpha}x})^2} \\ &= -\frac{8\sqrt{\alpha}}{\beta^2} \frac{1}{1 + e^{2\sqrt{\alpha}x}} \Big|_{-\infty}^{+\infty} \end{aligned}$$

$$= \frac{8\sqrt{\alpha}}{\beta^2}. \tag{2.29}$$

Due to the Lorentz symmetry of the system the energy of a moving soliton equals

$$E = \frac{8\sqrt{\alpha}}{\beta^2}\gamma. \tag{2.30}$$

Owing to techniques like the inverse scattering method, the Hirota method, the Bäcklund transformation, the Darboux transformation and the Painlev analysis there are many solutions of the sine–Gordon model known (see Refs. [63, 64, 65, 61]). In the following we discuss only superficially the best-known solutions and the Bäcklund transformation. The solution representing a *solution–antisoliton* state (see Refs. [56, 54, 76]) is given by

$$\varphi_{SA}(x) = \frac{4}{\beta} \arctan\left(\frac{\sinh(\sqrt{\alpha}\gamma vt)}{v\cosh(\sqrt{\alpha}\gamma x)}\right). \tag{2.31}$$

It is depicted in Fig. (2.3). Using the sine–Gordon equation it can easily be verified that this is an exact solution. The asymptotic behaviour in the past

$$\lim_{t\to-\infty} \varphi_{SA}(x) = \frac{4}{\beta}\arctan\exp\left(\sqrt{\alpha}\gamma(x+v(t+\delta/2))\right) - \frac{4}{\beta}\arctan\exp\left(\sqrt{\alpha}\gamma(x-v(t+\delta/2))\right)$$
$$= \varphi_S\left(\sqrt{\alpha}\gamma(x+v(t+\delta/2))\right) + \varphi_A\left(\sqrt{\alpha}\gamma(x-v(t+\delta/2))\right), \tag{2.32}$$

and future

$$\lim_{t\to+\infty} \varphi_{SA}(x) = \varphi_S\left(\sqrt{\alpha}\gamma(x+v(t-\delta/2))\right) + \varphi_A\left(\sqrt{\alpha}\gamma(x-v(t-\delta/2))\right), \tag{2.33}$$

where $\delta = \ln v/(v\gamma^2)$, reveals the interpretation as soliton–antisoliton scattering. At $t\to-\infty$ an antisoliton approaches from $x\to-\infty$ with velocity $+v$ to a soliton which propagates from $x\to+\infty$ with velocity $-v$. In the region around $t=0, x=0$ they overlap, interact and leave, the antisoliton to $x\to+\infty$ and the soliton to $x\to-\infty$ for $t\to+\infty$. The phase shift δ is the result of their interaction. Since it is negative the interaction between the (anti-)soliton pair is attractive.

The solution representing a *soliton–soliton* state is given by by (see Fig. (2.4))

$$\varphi_{SS}(x) = \frac{4}{\beta}\arctan\left(\frac{v\sinh(\sqrt{\alpha}\gamma x)}{\cosh(\sqrt{\alpha}\gamma vt)}\right). \tag{2.34}$$

Here two initially separated solitons approach each other, backward scattering occurs and they leave in directions opposite to their initial ones. If we identify the two solitons modulo $2\pi n/\beta$, $n\in\mathbb{N}$ one could alternatively interpret this solution as forward scattering of two solitons. The phase shift δ occurs again but in this case the interaction is repulsive.

Using Eq. (A.4) a *antisoliton–antisoliton* state $\varphi_{AA}(x)$ is obtained

$$\varphi_{AA}(x) = -\varphi_{SS}(x). \tag{2.35}$$

2.1. Introduction to the "classical" sine–Gordon model 21

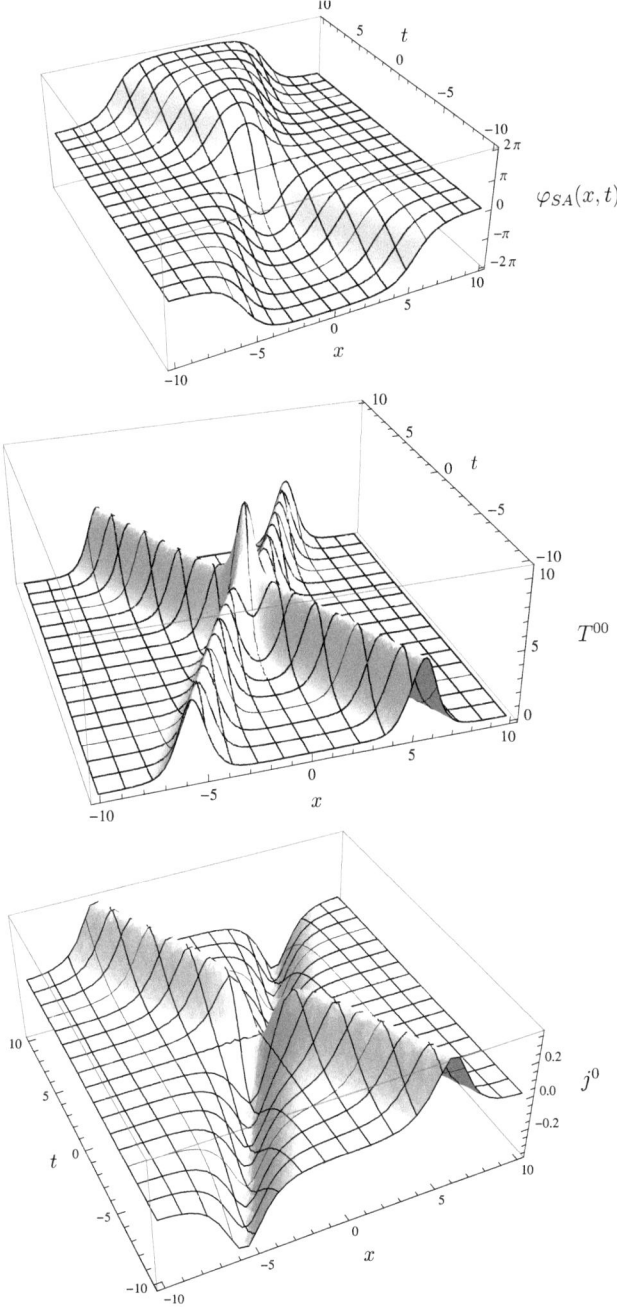

Figure 2.3.: The "soliton–antisoliton" state: the profile $\varphi_{SA}(x,t)$ with parameters $\alpha = \beta = 1$ and $v = 0.5$ (*top*), the energy density T^{00} (*middle*) and the current density j^0 (*below*).

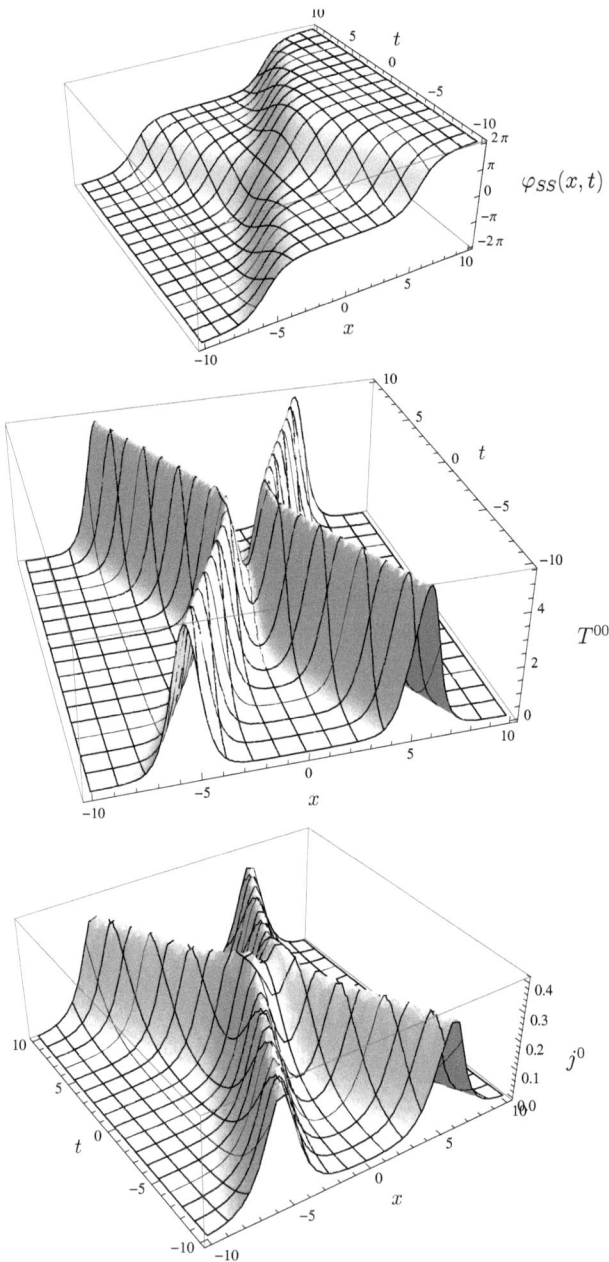

Figure 2.4.: The soliton–soliton state: the profile $\varphi_{SS}(x,t)$ with parameters $\alpha = \beta = 1$ and $v = 0.5$ (*top*), the energy density T^{00} (*middle*) and the current density j^0 (*below*).

Another family of solutions is called *breathers*, which is given by (see Fig. (2.5))

$$\varphi_B(x) = \frac{4}{\beta} \arctan\left(\frac{\sin(\sqrt{\alpha}ut/\sqrt{1+u^2})}{u\cosh(\sqrt{\alpha}x/\sqrt{1+u^2})}\right). \quad (2.36)$$

This solution is obtained by setting $v = iu$ in $\varphi_{SA}(x)$ still yielding a real function. This solution describes a soliton–antisoliton bound state, with period $\tau = 2\pi\sqrt{1+u^2}/(\sqrt{\alpha}u)$. The pair separates to some finite distance and collapses again due to the attractive interaction yielding an oscillating movement. Furthermore, there are no breather solutions for values of the coupling constant $\beta^2 > 8\pi$, since the number n of breathers depends on the value of the coupling constant β^2, i.e. $n = 1, 2, \ldots, 8\pi/\beta^2$ (see Ref. [46]).

Setting $v = iu$ in the solution $\varphi_{AA}(x)$ or $\varphi_{SS}(x)$ yields a complex function, which we exclude for the real sine–Gordon system. This agrees well with the non-existence of bound states for a system with repulsive interaction.

Finally, we give a superficially review on the *Bäcklund transformation* (for details see Ref. [61]), which is a very elegant method to obtain new solutions from known ones. In light-like coordinates $\sigma = \frac{1}{2}(x+t)$ and $\rho = \frac{1}{2}(x-t)$ the sine–Gordon equation reads

$$\frac{\partial}{\partial\sigma}\frac{\partial}{\partial\rho}\varphi(x) - \frac{\alpha}{\beta}\sin\beta\varphi(x) = 0. \quad (2.37)$$

The set of equations

$$\frac{1}{2}\frac{\partial}{\partial\sigma}(\varphi_1(x) - \varphi_0(x)) = \lambda\frac{\sqrt{\alpha}}{\beta}\sin\left(\frac{\beta}{2}(\varphi_1(x) + \varphi_0(x))\right), \quad (2.38)$$

$$\frac{1}{2}\frac{\partial}{\partial\rho}(\varphi_1(x) + \varphi_0(x)) = \frac{1}{\lambda}\frac{\sqrt{\alpha}}{\beta}\sin\left(\frac{\beta}{2}(\varphi_1(x) - \varphi_0(x))\right), \quad (2.39)$$

where λ is a real parameter, yields on operating with $\frac{\partial}{\partial\rho}$ on Eq. (2.38) and using Eq. (2.39)

$$\frac{1}{2}\frac{\partial}{\partial\sigma}\frac{\partial}{\partial\rho}(\varphi_1(x) - \varphi_0(x)) = \frac{\alpha}{\beta}\cos\left(\frac{\beta}{2}(\varphi_1(x) + \varphi_0(x))\right)\sin\left(\frac{\beta}{2}(\varphi_1(x) - \varphi_0(x))\right)$$

$$= \frac{1}{2}\frac{\alpha}{\beta}\sin(\beta\varphi_1(x)) - \frac{1}{2}\frac{\alpha}{\beta}\sin(\beta\varphi_0(x)). \quad (2.40)$$

From this one can read off that if $\varphi_0(x)$ is solution to the sine–Gordon equation, so must be $\varphi_1(x)$. For a given solution Eqs. (2.38) and (2.39) yield another solution. It is easy to verify that for $\varphi_0(x) = 0$ the soliton solution Eq. (2.26) is obtained for $\varphi_1(x)$.

2.2. Introduction to the "quantum" sine–Gordon model

According to Coleman [30] the sine–Gordon model does not exist for values of the coupling constant $\beta^2 > 8\pi$, since in this case the energy density of the vacuum state is unbounded from below. Such a result was supported by Dashen et al. [71, 72], who found that there is a finite quantum correction to the mass of a soliton. This finite correction introduces a

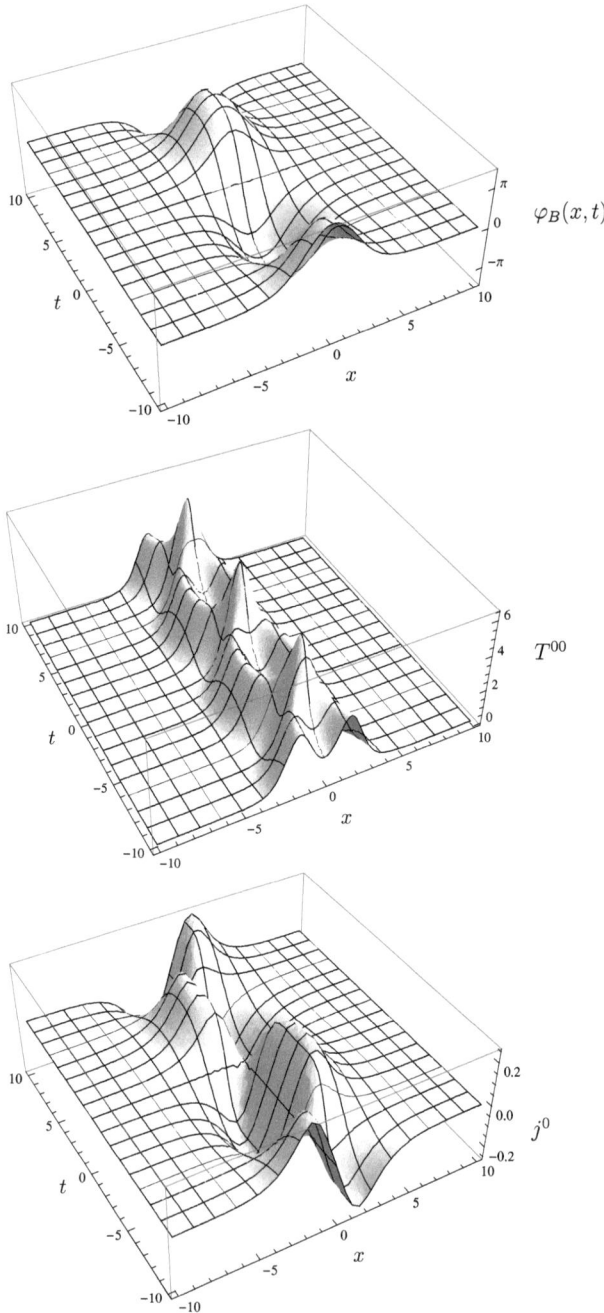

Figure 2.5.: The "breather" states: the profile $\varphi_B(x,t)$ with parameters $\alpha = \beta = 1$ and $u = 0.5$ (*top*), the energy density T^{00} (*middle*) and the current density j^0 (*below*).

2.2. Introduction to the "quantum" sine–Gordon model

singularity in the sine–Gordon model at the value of the coupling constant $\beta^2 = 8\pi$.

According to Zamolodchikov and Zamolodchikov [75] the appearance of such a singularity at $\beta^2 = 8\pi$ is a consequence of the regularisation and renormalisation procedure. Confirm of this assertion was given by Faber and Ivanov [46]. Furthermore, Amit et al. [36] have found that the coupling constant $\beta^2 = 8\pi$ defines the critical point of the Kosterlitz–Thouless phase transition. For $\beta^2 < 8\pi$ there is no long-range parameter, whereas for $\beta^2 > 8\pi$ there is a non-vanishing long-range parameter. Amit et al. also showed that in the vicinity of the coupling constant $\beta^2 \approx 8\pi$ the sine–Gordon model is renormalisable and regular. Since Amit et al. used non-standard approaches for the analysis of these problems, the confirmation of these results within the standard procedure of renormalisation and regularisation is meaningful. The renormalisation and regularisation of the sine–Gordon model within the standard procedure is the subject of this chapter.

In the following the complete renormalisability of the sine–Gordon model is shown by the example of the two-point causal Green function up to second order in α and all orders in β^2. First we rewrite the Lagrangian in terms of bare values. Since, as will be seen, the field $\varphi(x)$ and the coupling constant β are not affected by quantum corrections only the mass parameter α has to be replaced by a cutoff dependent bare value

$$\mathcal{L}(x) = \frac{1}{2}\partial_\mu\varphi(x)\partial^\mu\varphi(x) + \frac{\alpha_0(\Lambda^2)}{\beta^2}\left(\cos\beta\varphi(x) - 1\right), \tag{2.41}$$

where $\alpha_0(\Lambda^2)$ is a dimensional bare coupling constant and Λ is an ultraviolet cut-off. In [46] it was shown that the coupling constant $\alpha_0(\Lambda^2)$ is multiplicatively renormalisable

$$\alpha_0(\Lambda^2) = Z_1(\alpha(M^2), \beta^2, M^2; \Lambda^2)\,\alpha(M^2), \tag{2.42}$$

where $Z_1 = Z_1(\alpha(M^2), \beta^2, M^2; \Lambda^2)$ is the renormalisation constant [46], [66]–[68] which depends on the normalisation scale M and $\alpha(M^2)$ is the renormalised coupling constant which depends on the normalisation scale M instead of the cutoff Λ. Using Eq. (2.42) the Lagrangian can be rewritten

$$\mathcal{L}(x) = \frac{1}{2}\partial_\mu\varphi(x)\partial^\mu\varphi(x) + \frac{\alpha(M^2)}{\beta^2}\left(\cos\beta\varphi(x) - 1\right) + (Z_1 - 1)\frac{\alpha(M^2)}{\beta^2}\left(\cos\beta\varphi(x) - 1\right)$$
$$= \frac{1}{2}\partial_\mu\varphi(x)\partial^\mu\varphi(x) + Z_1\frac{\alpha(M^2)}{\beta^2}\left(\cos\beta\varphi(x) - 1\right). \tag{2.43}$$

In Ref. [46] the renormalisation constant $Z_1(\alpha(M^2), \beta^2, M^2; \Lambda)$ was found to be

$$Z_1(\alpha(M^2), \beta^2, M^2; \Lambda^2) = \left(\frac{\Lambda^2}{M^2}\right)^{\beta^2/8\pi}. \tag{2.44}$$

This result is non-perturbative and thus valid to all orders of perturbation theory. Since Z_1 does not depend on $\alpha(M^2)$, we write $Z_1 = Z_1(\beta^2, M^2; \Lambda^2)$ and abbreviate α for the mass parameter with dependence on M^2, further on.

For the investigation of the renormalisability with respect to quantum fluctuations around the trivial vacuum we expand the Lagrangian Eq. (2.43) in powers of $\varphi(x)$ as follows

$$\mathcal{L}(x) = \frac{1}{2}\partial_\mu\varphi(x)\partial^\mu\varphi(x) - \frac{\alpha}{2}\varphi^2(x) + \mathcal{L}_i(x)\,, \tag{2.45}$$

with the self–interaction part $\mathcal{L}_i(x)$

$$\mathcal{L}_i(x) = \alpha\sum_{n=2}^{\infty}\frac{(-1)^n}{(2n)!}\beta^{2(n-1)}\varphi^{2n}(x) + (Z_1 - 1)\alpha\sum_{n=1}^{\infty}\frac{(-1)^n}{(2n)!}\beta^{2(n-1)}\varphi^{2n}(x)\,. \tag{2.46}$$

This split in a free and interacting Lagrangian implies that the free propagator acquires a squared mass α. In principle the part proportional to $\varphi^2(x)$ of the full Lagrangian can be split arbitrarily between the free and interacting Lagrangian, hence determining what is called free propagation and what is accounted for as interaction. The free causal propagator reads with our split

$$-i\,\Delta_F(x-y) = \langle 0|\mathrm{T}(\varphi(x)\varphi(y))|0\rangle = \int\frac{d^2k}{(2\pi)^2 i}\frac{e^{-i\,k\cdot(x-y)}}{\alpha - k^2 - i\,0}\,. \tag{2.47}$$

With a Euclidean 2–dimensional momentum cut–off Λ the propagator equals at $x=y$ (see Appendix A.2.2)

$$-i\,\Delta_F(0) = \frac{1}{4\pi}\ln\left(\frac{\Lambda^2}{\alpha}\right). \tag{2.48}$$

2.3. Power counting and renormalisation of the sine–Gordon model

The general analysis of renormalisability of a quantum field theory starts by making use of the concept of the superficial degree of divergence of Feynman momentum integrals based on dimensional considerations [66]–[68] and by determining the dimension of the coupling constants.

The *superficial degree of divergence* D of a Feynman diagram is defined as

$$\begin{aligned}D &\equiv \text{(power of momenta in numerator)} - \text{(power of momenta in denominator)}\\ &= 2L - 2P\,,\end{aligned} \tag{2.49}$$

where L is the number of loops and P the number of propagators appearing in the diagram (as propagators we understand in the following only internal lines of a diagram and not the lines attached to external fields). The reason for this is clearly that each loop leads to an integration over the internal momentum of the loop, which contributes with a factor of two in the numerator, since space–time dimension is two. Each propagator contributes with a factor two in the denominator since the equation of motion is of second order. Naively, we expect a diagram with superficial degree of divergence D to scale as Λ^D for a cutoff Λ and thus diverging for $D > 0$, diverging like $\log\Lambda$ for $D = 0$ and converging to a finite

2.3. Power counting and renormalisation of the sine–Gordon model

value for $D < 0$. Unfortunately, this expectation is often wrong since if a diagram contains a diverging subdiagram the actual divergence may be worse than the superficial one. If symmetries like the Ward identity cause certain terms to cancel the actual divergence may be reduced or eliminated. Nevertheless, D is a useful quantity since according to *Weinberg's theorem* [69, 66, 70] a Feynman diagram converges if the superficial degree of divergence of the diagram and all its subdiagrams is negative.

The following relation holds

$$L = P - V + 1, \qquad (2.50)$$

where V is the number of vertices of a diagram. The validity of this equation can be seen as follows. Starting with some arbitrary diagram we see that the additional inclusion of a propagator between two vertices, such that the number of vertices is unchanged, increases the number of loops by one since the added propagator splits exactly one loop in two. Therefore, we have $L = P + C$ with C being some constant. Now suppose a diagram consisting of a loop with external legs on it. Adding one further external leg does not change the number of loops, it is still one, but the further leg splits one propagator in two and adds one vertex, thus increasing the number of propagators and vertices by one. But this means that $L = P - V + D$ with D being some constant. This constant is evaluated by counting loops, propagators and vertices of a given diagram. For example, the two point diagram of second order of a scalar ϕ^3 theory has one loop, two propagators and two vertices. Thus we are lead to Eq. (2.50), which itself leads to

$$D = 2 - 2V. \qquad (2.51)$$

From this we conclude that the most divergent diagram is the "vacuum bubble" which is one loop without external legs. Since this diagram has no subdiagrams and there are no appropriate symmetries the actual and superficial degrees of divergence coincide and the diagram diverges quadratically ($L = 1$, $P = 0$, $V = 0$, $D = 2$). However, the "vacuum bubble" does not contribute since it is not connected to the external legs. The second most divergent diagrams have $L = n$, $P = n$, $V = 1$, $D = 0$, $n \in \mathbb{N}$ or contain such graphs as subdiagrams and are logarithmically divergent. These diagrams contain at least one closed loop consisting of a single internal line and are sometimes referred to as tadpole diagrams. All connected diagrams which do not contain tadpoles are convergent since to each loop in all subdiagrams there are at least two propagators yielding $D < 0$ for the graph and all its subdiagrams, which results in a convergent Feynman diagram according to Weinberg's theorem. Furthermore, it is well known (see Ref. [30]) that for all scalar field theories in two dimensions with non-derivative interactions the only divergent contributions come from tadpole diagrams. But these are exactly the diagrams which are cancelled if the interaction operator is replaced by its normal-ordered form. Since a tadpole just gives a contribution to the mass of a particle, normal-ordering only means that these mass corrections are already taken into account in the mass parameter which appears in the free propagator. This corresponds to a shift in the energy contributions of the interactions. Instead of normal-ordering, the logarithmic divergences can also be absorbed by the coupling constant α.

In general all diverging quantum field theories can be divided in three classes (see e.g. Ref. [67])

- *Super-Renormalisable Theories:* There are only a finite number of Feynman diagrams which are superficially divergent and the coupling constant has positive mass dimension,

- *Renormalisable Theories:* There are only a finite number of amplitudes which are superficially divergent but divergences occur at all orders in perturbation theory and the coupling constant is dimensionless,

- *Non-Renormalisable Theories:* All amplitudes are divergent from some order in perturbation upwards and the coupling constant has negative mass dimension.

Since the coupling constant α has mass dimension two and β is dimensionless we conclude that according to this definition the sine–Gordon theory is a super-renormalisable theory.

2.4. Renormalisation of the causal two–point Green function

The generating functional of Green functions reads

$$Z[J] = \int \mathcal{D}\varphi \exp\left\{i \int d^2z \left[\mathcal{L}(z) + \varphi(z)J(z)\right]\right\}$$
$$= \int \mathcal{D}\varphi \exp\left\{i \int d^2z \left[\frac{1}{2}\left(\partial_\mu \varphi(z)\partial^\mu \varphi(z) - \alpha\,\varphi^2(z)\right) + \mathcal{L}_i(z) + \varphi(z)J(z)\right]\right\}, \quad (2.52)$$

where $J(z)$ is the external source of the sine–Gordon field $\varphi(z)$ and the normalisation $Z[0] = 1$ is understood. Functional derivation yields the causal two–point Green function

$$G(x,y) = \frac{1}{i}\frac{\delta}{\delta J(x)}\frac{1}{i}\frac{\delta}{\delta J(y)} Z[J]\Big|_{J=0}$$
$$= \int \mathcal{D}\varphi\, \varphi(x)\varphi(y) \exp\left\{i \int d^2z\, \mathcal{L}_i(z)\right\} \exp\left\{\frac{i}{2}\int d^2z \left[\partial_\mu \varphi(z)\partial^\mu \varphi(z) - \alpha\,\varphi^2(z)\right]\right\}.$$
(2.53)

In terms of the vacuum expectation value of a time–ordered operator product the Green function reads (see e.g. [74])

$$G(x,y) = \langle 0|\mathrm{T}\left(\varphi(x)\varphi(y) \exp\left\{i \int d^2z\, \mathcal{L}_i(z)\right\}\right)|0\rangle_c, \quad (2.54)$$

where the index c refers to the connected part, $\varphi(x)$ is the free sine–Gordon field operator with mass α and the causal two–point Green function $\Delta_F(x-y)$ is defined in Eq. (2.47).

2.4. Renormalisation of the causal two–point Green function

In the momentum representation the Green function Eq. (2.54) is

$$\tilde{G}(p) = \int d^2z\, e^{+ip\cdot z} G(z,0)$$
$$= \int d^2z\, e^{+ip\cdot z} \langle 0|T\Big(\varphi(z)\varphi(0)\exp\Big\{i\int d^2y\,\mathcal{L}_i(y)\Big\}\Big)|0\rangle_c. \tag{2.55}$$

Applying standard perturbation theory yields

$$G(x,y) = \sum_{n=0}^{\infty} G^{(n)}(x,y), \tag{2.56}$$

where

$$G^{(n)}(x,y) = \frac{i^n}{n!}\int d^2z_1\ldots\int d^2z_n\, \langle 0|T(\varphi(x)\varphi(y)\mathcal{L}_i(z_1)\ldots\mathcal{L}_i(z_n))|0\rangle_c. \tag{2.57}$$

In terms of α and β the n-th order term in Eq. (2.57) corresponds to the sum of all Feynman diagrams of all orders in β^2 and n-th order in α. To zeroth order the Green function $G^{(0)}(x,y)$ is $-i$ times the Feynman propagator Eq. (2.47). In the momentum representation we find

$$\tilde{G}^{(n)}(p) = \int d^2z\, e^{+ip\cdot z} G^{(n)}(z,0)$$
$$= \frac{i^n}{n!}\int d^2z\, e^{+ip\cdot z}\int d^2y_1\ldots\int d^2y_n\, \langle 0|T(\varphi(z)\varphi(0)\mathcal{L}_i(y_1)\ldots\mathcal{L}_i(y_n))|0\rangle_c. \tag{2.58}$$

Next we evaluate the quantum corrections up to second order in α and to all orders in β^2.

2.4.1. Expansion to first order in α and to all orders in β^2

Using Eq. (2.47) we find for the correction to first order in α and all orders in β^2

$$G^{(1)}(x,y) = i\int d^2z\, \langle 0|T(\varphi(x)\varphi(y)\mathcal{L}_i(z))|0\rangle_c$$
$$= i\alpha\sum_{m=2}^{\infty}\frac{(-1)^m}{(2m)!}\beta^{2(m-1)}\int d^2z\, \langle 0|T(\varphi(x)\varphi(y)\varphi^{2m}(z))|0\rangle_c$$
$$+ i\alpha(Z_1-1)\sum_{m=1}^{\infty}\frac{(-1)^m}{(2m)!}\beta^{2(m-1)}\int d^2z\, \langle 0|T(\varphi(x)\varphi(y)\varphi^{2m}(z))|0\rangle_c. \tag{2.59}$$

According to Wick's theorem the time-ordered product equals the sum of all permutations of the completely contracted expressions. Each contraction yields a propagator Eq. (2.47) with appropriate argument. Since we are evaluating the connected contribution, one Wick contraction combines the field located at x with one of the $2m$ fields located at z yielding a factor $-i\Delta_F(x-z)$. Another contraction combines the field at y with one of the, now only, $2m-1$ fields at z yielding $-i\Delta_F(y-z)$. Now to the contractions of the remaining fields at z. Lets consider all contractions split in a beginning A and an ending B. Since we are left with $2m-2$ fields located at z we have $2m-2$ fields for the first A, the first B has only $2m-3$ fields, the second A only $2m-4$ and so on. This yields a factor of $-i(2m-2)!\Delta_F(0)$. Since we have divided each contraction in beginning and ending we have obviously done some over-counting, because each field was contracted once by a beginning and once by and ending. But we can get rid off of this over-counting by dividing by a factor of 2 for all contractions between fields at z. Since these are $(2m - 2)/2$ contractions the contractions of the fields at z yield a factor of $-i(2m-2)!/(2^{m-1})\Delta_F(0)$. So we arrive at the result

$$\langle 0|T(\varphi(x)\varphi(y)\varphi^{2m}(z))|0\rangle_c = \frac{(2m)!}{2^{m-1}}[-i\Delta_F(x-z)][-i\Delta_F(y-z)][-i\Delta_F(0)]^{m-1}, \qquad (2.60)$$

respectively

$$\begin{aligned}
G^{(1)}(x,y) &= -i\alpha \sum_{m=2}^{\infty} \frac{\beta^{2(m-1)}}{2^{m-1}} [i\Delta_F(0)]^{m-1}(-i)^2 \int d^2z\, \Delta_F(x-z)\Delta_F(y-z) \\
&\quad - i\alpha(Z_1 - 1)\sum_{m=1}^{\infty} \frac{\beta^{2(m-1)}}{2^{m-1}} [i\Delta_F(0)]^{m-1}(-i)^2 \int d^2z\, \Delta_F(x-z)\Delta_F(y-z) \\
&= -i\alpha\left(\exp\left\{\frac{1}{2}\beta^2 i\Delta_F(0)\right\} - 1\right)(-i)^2 \int d^2z\, \Delta_F(x-z)\Delta_F(y-z) \\
&\quad - i\alpha(Z_1 - 1)\exp\left\{\frac{1}{2}\beta^2 i\Delta_F(0)\right\}(-i)^2 \int d^2z\, \Delta_F(x-z)\Delta_F(y-z) \\
&= i\alpha\left(1 - Z_1 \exp\left\{\frac{1}{2}\beta^2 i\Delta_F(0)\right\}\right)(-i)^2 \int d^2z\, \Delta_F(x-z)\Delta_F(y-z). \qquad (2.61)
\end{aligned}$$

Using Z_1 (see Eq. (2.44)) and the definition of the Green function for vanishing argument Eq. (2.48) we have succeeded in absorbing all divergences by the bare coupling constant $\alpha_0(\Lambda^2)$ and thus replacing the cutoff dependence in favour of a renormalisation point dependence M^2

$$G^{(1)}(x,y) = i\alpha\left(1 - \left(\frac{\alpha}{M^2}\right)^{\beta^2/8\pi}\right)(-i)^2 \int d^2z\, \Delta_F(x-z)\Delta_F(y-z). \qquad (2.62)$$

Adding the zeroth order contribution we find for the renormalised causal two–point Green function of the sine–Gordon field, defined to first order in α and to all orders in β^2

$$G(x,y) = -i\Delta_F(x-y) + i\alpha\left(1 - \left(\frac{\alpha}{M^2}\right)^{\beta^2/8\pi}\right)(-i)^2 \int d^2z\, \Delta_F(x-z)\Delta_F(y-z). \qquad (2.63)$$

In the momentum representation the two–point Green function reads

$$\tilde{G}(p) = \frac{-i}{\alpha - p^2} + \frac{-i}{\alpha - p^2}\, i\alpha\left(1 - \left(\frac{\alpha}{M^2}\right)^{\beta^2/8\pi}\right)\frac{-i}{\alpha - p^2}. \qquad (2.64)$$

The second term gives a correction to the mass of the sine–Gordon field. In order to evaluate the mass correction, we compare the relation

$$\begin{aligned}
\frac{-i}{\alpha + \delta\alpha - p^2} &= \frac{-i}{(\alpha - p^2)\left(1 + \frac{\delta\alpha}{\alpha - p^2}\right)} \\
&= \frac{-i}{\alpha - p^2} \sum_{n=0}^{\infty} \left(\frac{\delta\alpha}{\alpha - p^2}\right)^n
\end{aligned}$$

2.4. Renormalisation of the causal two–point Green function

$$= \frac{-i}{\alpha - p^2} - \frac{-i}{\alpha - p^2} i\delta\alpha \frac{-i}{\alpha - p^2} + \mathcal{O}(\delta\alpha^2), \tag{2.65}$$

with Eq. (2.64) and find for the mass correction to first order in α and all orders in β^2

$$\delta\alpha = -\alpha\left(1 - \left(\frac{\alpha}{M^2}\right)^{\tilde{\beta}^2/8\pi}\right). \tag{2.66}$$

Introducing the corrected mass

$$\alpha_{ph} = \alpha + \delta\alpha$$
$$= \alpha\left(\frac{\alpha}{M^2}\right)^{\tilde{\beta}^2/8\pi}, \tag{2.67}$$

the two–point Green function, calculated to first order in α and to all orders in β^2 can be written as

$$\tilde{G}(p) = \frac{-i}{\alpha_{ph} - p^2}. \tag{2.68}$$

Inverting Eq. (2.67) yields α in terms of M and α_{ph}

$$\alpha = \alpha_{ph}\left(\frac{M^2}{\alpha_{ph}}\right)^{\tilde{\beta}^2/8\pi}, \qquad \tilde{\beta}^2 = \frac{\beta^2}{1 + \frac{\beta^2}{8\pi}}. \tag{2.69}$$

Due to Eq. (2.68) the effective Lagrangian to first order in α and all orders in β^2 reads

$$\mathcal{L}_{\text{eff}}(x) = \frac{1}{2}\partial_\mu\varphi(x)\partial^\mu\varphi(x) + \frac{\alpha}{\beta^2}\left(\frac{\alpha}{M^2}\right)^{\tilde{\beta}^2/8\pi}(\cos\beta\varphi(x) - 1)$$
$$= \frac{1}{2}\partial_\mu\varphi(x)\partial^\mu\varphi(x) + \frac{\alpha_{ph}}{\beta^2}(\cos\beta\varphi(x) - 1). \tag{2.70}$$

Furthermore, we want to remark that if we make a renormalisation at the scale $M^2 = \alpha_{ph}$ we find, using Eqs. (2.66) and (2.69), that $\delta\alpha = 0$ and $\alpha_{ph} = \alpha$. This means that corrections to the mass do not appear to first order in α. They appear only at higher orders.

2.4.2. Expansion to second order in α and to all orders in β^2

The second order correction to the two–point Green function is given by

$$G^{(2)}(x,y) = -\frac{1}{2}\iint d^2y_1 d^2y_2 \,\langle 0|T(\varphi(x)\varphi(y)\mathcal{L}_i(y_1)\mathcal{L}_i(y_2))|0\rangle_c. \tag{2.71}$$

The calculation is rather cumbersome but runs parallel to the first order correction. It is performed in detail in Appendix B.1.1. The result reads

$$G^{(2)}(x,y) = \left(\alpha Z_1 \exp\left\{\frac{\beta^2}{2}i\Delta_F(0)\right\}\right)^2 \iint d^2z_1 d^2z_2\, [-i\Delta_F(x - z_1)]$$

$$\times \left(\cosh[-\beta^2 i\Delta_F(z_1 - z_2)] - 1 - \frac{1}{2}\beta^4 [-i\Delta_F(z_1 - z_2)]^2 \right) \frac{1}{\beta^2} [-i\Delta_F(z_1 - y)]$$

$$- \left[\alpha Z_1 \exp\left\{ \frac{\beta^2}{2} i\Delta_F(0) \right\} \right]^2 \iint d^2z_1 d^2z_2 \, [-i\Delta_F(x - z_1)]$$

$$\times \left(\sinh[-\beta^2 i\Delta_F(z_1 - z_2)] - \beta^2 [-i\Delta_F(z_1 - z_2)] \right) \frac{1}{\beta^2} [-i\Delta_F(z_2 - y)]. \quad (2.72)$$

In the momentum representation the details of the calculation are stated in Appendix B.1.2. The result reads

$$\tilde{G}^{(2)}(p) = i\alpha_{ph}^2 \left(\frac{-i}{\alpha_{ph} - p^2} \right)^2 \Bigg\{ \sum_{m=2}^{\infty} \frac{\beta^{4m-4}}{(2m-1)!} \int \frac{d^2k_1}{(2\pi)^2 i} \frac{1}{\alpha_{ph} - k_1^2} \int \frac{d^2k_2}{(2\pi)^2 i} \frac{1}{\alpha_{ph} - k_2^2} \cdots$$

$$\times \int \frac{d^2k_{2m-2}}{(2\pi)^2 i} \frac{1}{\alpha_{ph} - k_{2m-2}^2} \frac{1}{\alpha_{ph} - (p - k_1 - k_2 - \ldots - k_{2m-2})^2}$$

$$- \sum_{m=2}^{\infty} \frac{\beta^{4m-2}}{(2m)!} \int \frac{d^2k_1}{(2\pi)^2 i} \frac{1}{\alpha_{ph} - k_1^2} \int \frac{d^2k_2}{(2\pi)^2 i} \frac{1}{\alpha_{ph} - k_2^2} \cdots$$

$$\times \int \frac{d^2k_{2m-1}}{(2\pi)^2 i} \frac{1}{\alpha_{ph} - k_{2m-1}^2} \frac{1}{\alpha_{ph} - (k_1 + k_2 + \ldots + k_{2m-1})^2} \Bigg\}. \quad (2.73)$$

2.5. Renormalisation of the massive sine–Gordon model

Now we investigate the renormalisation of the two–point Green function for the sine–Gordon model with a mass term added [31]–[45], [47]. The bare Lagrangian becomes (see Ref. [36])

$$\mathcal{L}(x) = \frac{1}{2} \partial_\mu \varphi(x) \partial^\mu \varphi(x) - \frac{1}{2} m_0^2(\Lambda^2) \varphi^2(x) + \frac{\alpha_0(\Lambda^2)}{\beta^2} (\cos \beta \varphi(x) - 1), \quad (2.74)$$

where $m_0(\Lambda^2)$ is the additional bare mass. The renormalised Lagrangian is

$$\mathcal{L}(x) = \frac{1}{2} \partial_\mu \varphi(x) \partial^\mu \varphi(x) - \frac{1}{2} m^2(M^2) \varphi^2(x) + \frac{\alpha}{\beta^2}(\cos \beta \varphi(x) - 1)$$

$$- \frac{1}{2} m^2(M^2) (Z_m - 1) \varphi^2(x) + (Z_1 - 1) \frac{\alpha}{\beta^2} (\cos \beta \varphi(x) - 1)$$

$$= \frac{1}{2} \partial_\mu \varphi(x) \partial^\mu \varphi(x) - \frac{1}{2} Z_m m^2(M^2) \varphi^2(x) + Z_1 \frac{\alpha}{\beta^2} (\cos \beta \varphi(x) - 1), \quad (2.75)$$

where Z_1 is defined by (2.44) and $Z_m = Z_m(\alpha, \beta^2, M^2; \Lambda^2)$ is the renormalisation constant of the mass

$$m(M^2) = Z_m^{-1/2}(\alpha, \beta^2, M^2; \Lambda^2) \, m_0(\Lambda^2). \quad (2.76)$$

For the analysis of the renormalisability we proceed analogous to the massless case and expand the Lagrangian Eq. (2.75) in powers of $\varphi(x)$. Defining the effective mass $\mu^2(M^2) = m^2(M^2) + \alpha(M^2)$ (the explicit dependence on M^2 will not be denoted further on) we find

$$\mathcal{L}(x) = \frac{1}{2} \partial_\mu \varphi(x) \partial^\mu \varphi(x) - \frac{\mu^2}{2} \varphi^2(x) + \mathcal{L}_i(x), \quad (2.77)$$

2.5. Renormalisation of the massive sine–Gordon model

with the interaction part of the Lagrangian

$$\mathcal{L}_i(x) = -\frac{1}{2}m^2\,(Z_m-1)\,\varphi^2(x) + \alpha \sum_{n=2}^{\infty} \frac{(-1)^n}{(2n)!} \beta^{2(n-1)} \varphi^{2n}(x)$$
$$+ (Z_1 - 1)\,\alpha \sum_{n=1}^{\infty} \frac{(-1)^n}{(2n)!} \beta^{2(n-1)} \varphi^{2n}(x)\,. \tag{2.78}$$

With this split between free part and interaction part the mass parameter appearing in the free propagator is μ. The Feynman propagator reads

$$-i\Delta_F^{(m)}(x-y) = \langle 0|\mathrm{T}(\varphi(x)\varphi(y))|0\rangle = \int \frac{d^2k}{(2\pi)^2 i}\,\frac{e^{-ik\cdot x}}{\mu^2 - k^2 - i0}\,. \tag{2.79}$$

At vanishing argument the propagator is equal to (see Appendix A.2.2)

$$-i\Delta_F^{(m)}(0) = \frac{1}{4\pi}\ln\left(\frac{\Lambda^2}{\mu^2}\right), \tag{2.80}$$

with the Euclidean cut-off Λ.

The calculation of the correction to first order in α and to all orders in β^2 runs parallel to that in Eq. (2.61)

$$G_{(m)}(x,y) = -i\Delta_F^{(m)}(x-y) + i\alpha\left(1 - Z_1\exp\left\{\frac{1}{2}\beta^2 i\Delta_F(0,\mu^2)\right\}\right)$$
$$\times (-i)^2 \int d^2z\,\Delta_F(x-z)\Delta_F(y-z)$$
$$+ (Z_m - 1)\,m^2(-i)^3 \int d^2z\,\Delta_F(x-z)\Delta_F(y-z)\,. \tag{2.81}$$

In the momentum representation this expression takes the form

$$\tilde{G}_{(m)}(p) = \frac{-i}{\mu^2 - p^2} + \frac{-i}{\mu^2 - p^2}\,i\alpha\left(1 - \left(\frac{\mu^2}{M^2}\right)^{\beta^2/8\pi}\right)\frac{-i}{\mu^2 - p^2}$$
$$- \frac{-i}{\mu^2 - p^2}\,i(Z_m - 1)\,m^2\,\frac{-i}{\mu^2 - p^2}\,. \tag{2.82}$$

Analogous to the massless case we find for the correction to the mass

$$\delta m^2 = -\alpha\left(1 - \left(\frac{\mu^2}{M^2}\right)^{\beta^2/8\pi}\right) + (Z_m - 1)\,m^2\,, \tag{2.83}$$

yielding

$$m_{ph}^2 = \mu^2 + \delta m^2$$
$$= \mu^2 - \alpha\left(1 - \left(\frac{m^2+\alpha}{M^2}\right)^{\beta^2/8\pi}\right) + (Z_m - 1)\,m^2$$
$$= m^2 + \alpha\left(\frac{m^2+\alpha}{M^2}\right)^{\beta^2/8\pi} + (Z_m - 1)\,m^2\,. \tag{2.84}$$

The first two terms are independent of the cut–off Λ. Therefore, the last term must vanish and we find $Z_m = 1 + O(\alpha^2)$ and

$$m_{ph}^2 = m^2 + \alpha \left(\frac{m^2 + \alpha}{M^2}\right)^{\beta^2/8\pi}. \tag{2.85}$$

The two–point Green function, calculated to first order in α and to all orders in β^2 reads

$$\tilde{G}_{(m)}(p) = \frac{-i}{m_{ph}^2 - p^2}. \tag{2.86}$$

The calculation of the second order correction to the two–point Green function runs parallel to the massless case and can be found in Appendices B.1.3 and B.1.4 in detail. One can show that $Z_m = 1$, therefore the mass parameter $m_0(\Lambda^2)$ is unrenormalisable and we find

$$m_{ph}^2 = m_0^2 + \alpha \left(\frac{m_0^2}{M^2}\right)^{\beta^2/8\pi}. \tag{2.87}$$

Independence of the physical mass of the massive sine–Gordon model field on the normalisation scale demands to set

$$\alpha_{ph} = \alpha \left(\frac{m_0^2}{M^2}\right)^{\beta^2/8\pi} \longrightarrow \alpha = \alpha_{ph} \left(\frac{M^2}{m_0^2}\right)^{\beta^2/8\pi}. \tag{2.88}$$

Setting $M = m_0$ we find the relation $\alpha(m_0^2) = \alpha_{ph}$.

2.6. Renormalisation of the massless sine–Gordon model around solitons

Here the renormalisation procedure expounded above is applied to quantum corrections around a topologically non-trivial vacuum (this means that the topological charge, defined in Eq. (2.14), of the vacuum satisfies $Q \neq 0$) – the soliton solution (see Refs. [71, 72] and [76]), which is called vacuum since a soliton at rest is the lowest energy configuration. First the field is split as $\varphi(x) = \varphi_{cl}(x) + \phi(x)$, where $\varphi_{cl}(x)$ is solution to the classical equation of motion

$$\Box \varphi_{cl}(x) + \frac{\alpha_0}{\beta} \sin \beta \varphi_{cl}(x) = 0, \tag{2.89}$$

and $\phi(x)$ fluctuates around this classical solution. From this perspective the quantisation performed in the preceding sections can be viewed as quantisation around a topologically trivial vacuum ($Q = 0$) with $\varphi_{cl}(x) = 0$. Substituting $\varphi(x) = \varphi_{cl}(x) + \phi(x)$ in the sine–Gordon Lagrangian yields

$$\frac{1}{2} \partial_\mu \varphi(x) \partial^\mu \varphi(x) + \frac{\alpha_0}{\beta^2} \left(\cos \beta \varphi(x) - 1\right) = \frac{1}{2} \partial_\mu \varphi_{cl}(x) \partial^\mu \varphi_{cl}(x) + \frac{\alpha_0}{\beta^2} \left(\cos \beta \varphi_{cl}(x) - 1\right)$$

2.6. Renormalisation of the massless sine–Gordon model around solitons

$$+ \frac{1}{2}\partial_\mu \phi(x)\partial^\mu \phi(x) - \frac{\alpha_0}{2}\phi(x)^2 \cos\beta\varphi_{c\ell}(x)$$
$$+ \mathcal{O}(\phi(x)^3), \tag{2.90}$$

where terms of order $\phi(x)$ vanish since $\varphi_{c\ell}(x)$ is solution to the equation of motion.

In this section we quantise around the soliton solution (see Eq. 2.26)

$$\varphi_{c\ell}(x) = \frac{4}{\beta}\arctan\exp(\sqrt{\alpha_0}\,\gamma(x - vt))$$
$$= \frac{4}{\beta}\arctan\exp(\sqrt{\alpha_0}\,\rho)$$
$$= \varphi_S(\rho), \tag{2.91}$$

where we have used the "comoving coordinate" $\rho = \gamma(x - vt)$. The corresponding "comoving time coordinate" is $\tau = \gamma(t - vx)$. In variables (τ, ρ) an infinitesimal 2-dimensional volume element is $d^2x = d\tau d\rho$ and the d'Alembert operator reads $\Box = \partial^2/\partial\tau^2 - \partial^2/\partial\rho^2$. Since (see Appendix B.2 for a derivation)

$$\cos\beta\varphi_S(x) = \cos(4\arctan\exp(\sqrt{\alpha_0}\,\rho))$$
$$= 1 - \frac{2}{\cosh^2(\sqrt{\alpha_0}\rho)}, \tag{2.92}$$

we find

$$\mathcal{L}[\varphi(\tau,\rho)] = \mathcal{L}[\varphi_S(\rho)] - \frac{1}{2}\phi(\tau,\rho)\left[\Box + \alpha_0 - \frac{2\alpha_0}{\cosh^2(\sqrt{\alpha_0}\rho)}\right]\phi(\tau,\rho) + \mathcal{O}(\phi(\tau,\rho)^3), \tag{2.93}$$

where $\mathcal{L}[\varphi_S(\rho)]$ corresponds to the second line in Eq. (2.90).

Substituting Eq. (2.93) in the partition function

$$Z = \int \mathcal{D}\varphi \exp\left\{i\int d^2x \left[\frac{1}{2}\partial_\mu\varphi(x)\partial^\mu\varphi(x) + \frac{\alpha_0}{\beta^2}(\cos\beta\varphi(x) - 1)\right]\right\}$$
$$= \int \mathcal{D}\varphi \exp\left\{i\int d^2x \,\mathcal{L}[\varphi(x)]\right\}, \tag{2.94}$$

we find

$$Z = \exp\left\{i\int d^2x \,\mathcal{L}[\varphi_S(\rho)]\right\}$$
$$\times \int \mathcal{D}\phi \exp\left\{-i\frac{1}{2}\int d\tau d\rho \,\phi(\tau,\rho)\left[\Box + \alpha_0 - \frac{2\alpha_0}{\cosh^2(\sqrt{\alpha_0}\rho)}\right]\phi(\tau,\rho)\right\}, \tag{2.95}$$

where all terms of order $\phi(\tau,\rho)^3$ and higher have been dropped since we will restrict to corrections of second order further on. We want to remark that α_0 should enter with the imaginary correction $\alpha_0 \to \alpha_0 - i0$, which is required by the convergence of the path integral (see e.g. Ref. [78]). The attracting "potential" $V(\rho) = \alpha_0 - \frac{2\alpha_0}{\cosh^2(\sqrt{\alpha_0}\rho)}$ of the quantum fluctuations is depicted in Fig. 2.6. Performing the path integration the partition function

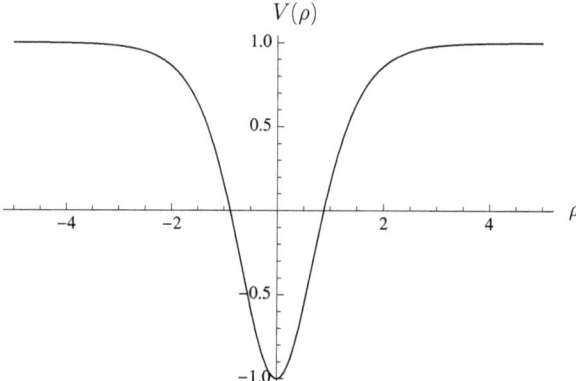

Figure 2.6.: Attractive "potential" $V(\rho) = \alpha_0 - \dfrac{2\alpha_0}{\cosh^2(\sqrt{\alpha_0}\rho)}$ for the quantum fluctuations of $\alpha = 1$.

Eq. (2.95) reads

$$Z = \exp\left\{i\int d^2x \left(\mathcal{L}[\varphi_S(\rho)] + \delta\mathcal{L}_{\text{eff}}[\varphi_S(\rho)]\right)\right\}, \qquad (2.96)$$

where

$$\exp\left\{i\int d^2x\, \delta\mathcal{L}_{\text{eff}}[\varphi_S(\rho)]\right\} = \exp\{i\delta S[\varphi_S(\rho)]\}$$
$$= \frac{1}{\sqrt{\text{Det}\left(\Box + \alpha_0 - \dfrac{2\alpha_0}{\cosh^2(\sqrt{\alpha_0}\rho)}\right)}}. \qquad (2.97)$$

Since we are interested in the correction to the action, which comes solely from the soliton, we have to evaluate the expression $\delta_* S[\varphi_S(\rho)] = \delta S[\varphi_S(\rho)] - \delta S[0]$, where the subtraction cancels contributions which are also present in the trivial vacuum $\varphi_{cl}(x) = 0$. So we find

$$\exp\{i\delta_* S[\varphi_S(\rho)]\} = \sqrt{\frac{\text{Det}(\Box + \alpha_0)}{\text{Det}\left(\Box + \alpha_0 - \dfrac{2\alpha_0}{\cosh^2(\sqrt{\alpha_0}\rho)}\right)}}$$
$$= \exp\left\{-\frac{1}{2}\sum_n \ln \lambda_n^{(S)} + \frac{1}{2}\sum_n \ln \lambda_n^{(V)}\right\}, \qquad (2.98)$$

where the first term in the exponential comes from $i\delta S[\varphi_S(\rho)]$, while the second term corresponds to the subtraction of the trivial vacuum contribution $i\delta S[0]$ and $\lambda_n^{(S)}$ are the eigenvalues of the equation

$$\left(\Box + \alpha_0 - \frac{2\alpha_0}{\cosh^2(\sqrt{\alpha_0}\rho)}\right)\phi_n(\tau,\rho) = \lambda_n^{(S)} \phi_n(\tau,\rho), \qquad (2.99)$$

2.6. Renormalisation of the massless sine–Gordon model around solitons

and $\phi_n(\tau, \rho)$ are the corresponding eigenfunctions. The quantum number n in Eq. (2.98) is partly discrete and has also a continuous part for the first term corresponding to the soliton, while the second term due to the vacuum is purely continuous. Therefore, the sum over n is understood as summation over the discrete values and as integration over the continuous ones.

Separating the time dependence and space dependence (see Ref. [77]) yields

$$\phi_n(\tau, \rho) = e^{-i\omega\tau} \psi_n(\rho), \qquad (2.100)$$

where $-\infty \leq \omega \leq +\infty$ and $\psi_n(\rho)$ is a complex function. Therefore, we have to take $\phi(\tau, \rho) = \mathcal{R}e\left(e^{-i\omega\tau}\psi(\rho)\right)$. One can also use complex eigenfunctions without loss of generality (see Ref. [72, 73, 76]).

Using Eq. (2.100) in Eq. (2.99) yields

$$\left(\frac{d^2}{d\rho^2} + k^2 + \frac{2\alpha_0}{\cosh^2(\sqrt{\alpha_0}\,\rho)}\right) \psi_n(\tau, \rho) = 0, \qquad (2.101)$$

where

$$k^2 = \lambda_n^{(S)} + \omega^2 - \alpha_0, \qquad (2.102)$$

is a spatial momentum $-\infty < k < +\infty$. For the eigenvalues $\lambda_n^{(S)}$ we find therefore

$$\lambda_n^{(S)} = \alpha_0 - \omega^2 + k^2. \qquad (2.103)$$

The solutions of Eq. (2.101) are derived in Appendix B.2 and read (see also Refs. [76, 72, 73])

$$\begin{aligned}
\psi_b(\rho) &= \sqrt{\frac{\sqrt{\alpha_0}}{2}} \frac{1}{\cosh(\sqrt{\alpha_0}\rho)}, \\
\psi_k(\rho) &= \frac{i}{\sqrt{2\pi}} \frac{-ik + \sqrt{\alpha_0}\,\tanh(\sqrt{\alpha_0}\rho)}{\sqrt{k^2 + \alpha_0}} e^{+ik\rho},
\end{aligned} \qquad (2.104)$$

respectively the time dependent fluctuating field is equal to

$$\begin{aligned}
\phi_{\omega b}(\tau, \rho) &= \frac{1}{\sqrt{2\pi}} e^{-i\omega\tau} \psi_b(\rho) = \frac{1}{\sqrt{2\pi}} \sqrt{\frac{\sqrt{\alpha_0}}{2}} \frac{1}{\cosh(\sqrt{\alpha_0}\rho)} e^{-i\omega\tau}, \\
\phi_{\omega k}(\tau, \rho) &= \frac{1}{\sqrt{2\pi}} e^{-i\omega\tau} \psi_k(\rho) = \frac{i}{2\pi} \frac{-ik + \sqrt{\alpha_0}\,\tanh(\sqrt{\alpha_0}\rho)}{\sqrt{k^2 + \alpha_0}} e^{-i\omega\tau + ik\rho}.
\end{aligned} \qquad (2.105)$$

The eigenvalues of the eigenfunctions $\psi_b(\rho)$, $\phi_b(\rho)$ read $\lambda_n^{(S)} = -\omega^2$ and $\psi_k(\rho)$ respectively $\phi_k(\rho)$ have eigenvalues $\lambda_n^{(S)} = \alpha_0 - \omega^2 + k^2$. From Eq. (2.105) we can read off the asymptotic region of the function $\phi_k(\rho)$ as

$$\lim_{\rho \to +\infty} \phi_{\omega k}(\tau, \rho) = \frac{1}{2\pi} e^{-i\omega\tau + ik\rho + i\frac{1}{2}\delta(k)},$$

$$\lim_{\rho \to -\infty} \phi_{\omega k}(\tau, \rho) = \frac{1}{2\pi} e^{-i\omega\tau + ik\rho - i\frac{1}{2}\delta(k)}, \qquad (2.106)$$

with the phase shift $\delta(k)$ (see also Ref. [76])

$$\delta(k) = 2 \arctan \frac{\sqrt{\alpha_0}}{k}. \qquad (2.107)$$

The corresponding solution in the vacuum sector is obtained by setting $\varphi_{cl} = 0$ in Eq. (2.90). The eigenvalue equation Eq. (2.99) reduces to the wave equation

$$(\Box + \alpha_0)\, \phi_n(\tau, \rho) = \lambda_n^{(V)}\, \phi_n(\tau, \rho), \qquad (2.108)$$

and after the separation Eq. (2.100) to

$$\left(\frac{d^2}{d\rho^2} + q^2\right) \psi_n(\tau, \rho) = 0, \qquad (2.109)$$

where

$$\lambda_n^{(V)} = \alpha_0 - \omega^2 + q^2. \qquad (2.110)$$

The obtained solution is

$$\phi_{\omega q}(\tau, \rho) = \frac{1}{\sqrt{2\pi}} e^{-i\omega\tau} \psi_q(\rho) = \frac{1}{2\pi} e^{-i\omega\tau + iq\rho}. \qquad (2.111)$$

In terms of these eigenfunctions we find for the effective action $\delta_* S[\varphi_S(\rho)]$

$$\begin{aligned}
\delta_* S[\varphi_S(\rho)] &= -\frac{1}{2} \int d^2x \int_{-\infty}^{+\infty} \frac{d\omega}{2\pi i} \int_{-\infty}^{+\infty} dk\, |\psi_k(x)|^2 \ln(\alpha_0 - \omega^2 + k^2) \\
&\quad - \frac{1}{2} \int d^2x \int_{-\infty}^{+\infty} \frac{d\omega}{2\pi i} |\psi_b(x)|^2 \ln(-\omega^2) \\
&\quad + \frac{1}{2} \int d^2x \int_{-\infty}^{+\infty} \frac{d\omega}{2\pi i} \int_{-\infty}^{+\infty} dq\, |\psi_q(x)|^2 \ln(\alpha_0 - \omega^2 + q^2) \\
&= -\frac{1}{2} \int d^2x \int_{-\infty}^{+\infty} \frac{d\omega}{2\pi i} \int_{-\infty}^{+\infty} \frac{dk}{2\pi} \frac{k^2 + \alpha_0 \tanh^2(\sqrt{\alpha_0}\rho)}{k^2 + \alpha_0} \ln(\alpha_0 - \omega^2 + k^2) \\
&\quad - \frac{1}{2} \int d^2x \int_{-\infty}^{+\infty} \frac{d\omega}{2\pi i} \frac{\sqrt{\alpha_0}}{2} \frac{1}{\cosh^2(\sqrt{\alpha_0}\rho)} \ln(-\omega^2) \\
&\quad + \frac{1}{2} \int d^2x \int \frac{d^2p}{(2\pi)^2 i} \ln(\alpha_0 - p^2),
\end{aligned} \qquad (2.112)$$

where we have introduced the 2-dimensional momentum p in the last line. Since the following relation holds

$$\frac{1}{\cosh^2(\sqrt{\alpha_0}\rho)} = \frac{1}{2}(1 - \cos\beta\varphi_S(\rho)) = \frac{\beta^2}{2\alpha_0} U[\varphi_S(\rho)], \qquad (2.113)$$

2.6. Renormalisation of the massless sine–Gordon model around solitons

we find for the corresponding effective Lagrangian

$$\delta_*\mathcal{L}_{\text{eff}}[\varphi_S(\rho)] = -\frac{\beta^2}{4}U[\varphi_S(\rho)]\int_{-\infty}^{+\infty}\frac{d\omega}{2\pi i}\int_{-\infty}^{+\infty}\frac{dk}{2\pi}\frac{1}{k^2+\alpha_0}\left[\ln(-\omega^2)-\ln(\alpha_0-\omega^2+k^2)\right]$$
$$-\frac{1}{2}\int_{-\infty}^{+\infty}\frac{d\omega}{2\pi i}\int_{-\infty}^{+\infty}\frac{dk}{2\pi}\ln(\alpha_0-\omega^2+k^2)+\frac{1}{2}\int\frac{d^2p}{(2\pi)^2i}\ln(\alpha_0-p^2)$$
$$= -\frac{\beta^2}{4}U[\varphi_S(\rho)]\int_{-\infty}^{+\infty}\frac{d\omega}{2\pi i}\int_{-\infty}^{+\infty}\frac{dk}{2\pi}\frac{1}{k^2+\alpha_0}\left[\ln(-\omega^2)-\ln(\alpha_0-\omega^2+k^2)\right]$$
$$= \frac{\beta^2}{2}U[\varphi_S(\rho)]\int_{-\infty}^{+\infty}\frac{dk}{2\pi}\int_{-\infty}^{+\infty}\frac{d\omega}{2\pi i}\frac{1}{\alpha_0-\omega^2+k^2-i0}, \quad (2.114)$$

where we have performed the partial integration with respect to ω in the last line and have used the integral representation

$$\frac{1}{2\sqrt{\alpha_0}} = \int_{-\infty}^{+\infty}\frac{dk}{2\pi}\frac{1}{k^2+\alpha_0}, \quad (2.115)$$

for the discrete mode $\lambda_n^{(S)} = -\omega^2$. Performing the ω integration yields

$$\delta_*\mathcal{L}_{\text{eff}}[\varphi_S(\rho)] = \frac{\beta^2}{2}U[\varphi_S(\rho)]\int_{-\infty}^{+\infty}\frac{dk}{4\pi}\frac{1}{\sqrt{\alpha_0+k^2}}$$
$$= -\frac{\beta^2}{4\sqrt{\alpha_0}}U[\varphi_S(\rho)]\int_{-\infty}^{+\infty}\frac{dk}{4\pi}\sqrt{\alpha_0+k^2}\,\frac{d\delta(k)}{dk}. \quad (2.116)$$

In order to investigate the renormalisability of the sine–Gordon and regularise in a covariant way we perform a Wick rotation $\omega \to i\omega$ in Eq. (2.114) and pass to Euclidean momentum space

$$\delta_*\mathcal{L}_{\text{eff}}[\varphi_S(\rho)] = \frac{\beta^2}{2}U[\varphi_S(\rho)]\int_{-\infty}^{+\infty}\frac{dk}{2\pi}\int_{-\infty}^{+\infty}\frac{d\omega}{2\pi i}\frac{1}{\alpha_0-\omega^2+k^2-i0}$$
$$= \frac{\beta^2}{8\pi}U[\varphi_S(\rho)]\ln\left(\frac{\Lambda^2}{\alpha_0}\right)$$
$$= -\frac{\alpha_0}{8\pi}\ln\left(\frac{\Lambda^2}{\alpha_0}\right)(\cos\beta\varphi_S(\rho)-1), \quad (2.117)$$

with the Euclidean cut–off Λ (see Ref. [46]). Adding the correction to the Lagrangian $\delta_*\mathcal{L}_{\text{eff}}[\varphi_S(\rho)]$ with the classical Lagrangian $\mathcal{L}[\varphi_S(\rho)]$ we find

$$\mathcal{L}_{\text{eff}}(x) = \frac{1}{2}\partial_\mu\varphi_S(\rho)\partial^\mu\varphi_S(\rho) + \frac{\alpha_0}{\beta^2}\left[1-\frac{\beta^2}{8\pi}\ln\left(\frac{\Lambda^2}{\alpha_0}\right)\right](\cos\beta\varphi_S(\rho)-1). \quad (2.118)$$

Renormalisation runs parallel to quantisation around the trivial vacuum

$$\alpha_0\left[1-\frac{\beta^2}{8\pi}\ln\left(\frac{\Lambda^2}{\alpha_0}\right)\right] = \alpha Z_1\left[1-\frac{\beta^2}{8\pi}\ln\left(\frac{\Lambda^2}{\alpha Z_1}\right)\right]$$
$$= \alpha\left[1+\frac{\beta^2}{8\pi}\ln\left(\frac{\Lambda^2}{M^2}\right)\right]\left[1-\frac{\beta^2}{8\pi}\ln\left(\frac{\Lambda^2}{\alpha}\right)\right]$$

$$= \alpha \left[1 + \frac{\beta^2}{8\pi} \ln\left(\frac{\alpha}{M^2}\right)\right], \tag{2.119}$$

where we have kept only terms of order $\mathcal{O}(\beta^2)$ and have used that (see Eq. (2.44))

$$Z_1 = 1 + \frac{\beta^2}{8\pi} \ln\left(\frac{\Lambda^2}{M^2}\right) + \mathcal{O}(\beta^4). \tag{2.120}$$

Thus we find for the renormalised effective Lagrangian

$$\begin{aligned}\mathcal{L}_{\text{eff}}(x) &= \frac{1}{2}\partial_\mu \varphi_S(\rho)\partial^\mu \varphi_S(\rho) + \frac{\alpha}{\beta^2}\left[1 + \frac{\beta^2}{8\pi}\ln\left(\frac{\alpha}{M^2}\right)\right](\cos\beta\varphi_S(\rho) - 1) \\ &= \frac{1}{2}\partial_\mu \varphi_S(\rho)\partial^\mu \varphi_S(\rho) + \frac{\alpha_{ph}}{\beta^2}(\cos\beta\varphi_S(\rho) - 1),\end{aligned} \tag{2.121}$$

where we have introduced the coupling constant α_{ph} (see Eq. (2.67))

$$\alpha_{ph} = \alpha\left[1 + \frac{\beta^2}{8\pi}\ln\left(\frac{\alpha}{M^2}\right)\right], \tag{2.122}$$

and have dropped all terms $\mathcal{O}(\beta^4)$.

One can see that the effective Lagrangian Eq. (2.121) coincides with the Lagrangian Eq. (2.70), renormalised by quantum corrections around the trivial vacuum.

We remark that the Gaussian fluctuations considered in this section are perturbative fluctuations of order $\mathcal{O}(\alpha\beta^2)$ valid for $\beta^2 \ll 8\pi$. Therefore they cannot be responsible for non–perturbative contributions to the soliton mass at $\beta^2 = 8\pi$.

2.7. Renormalisation of the soliton mass by Gaussian quantum corrections in continuous space–time

In this section we evaluate the mass of a quantum soliton

$$M_s = \frac{8\sqrt{\alpha_0}}{\beta^2} + \Delta M_s, \tag{2.123}$$

where the first part is the classical soliton mass Eq. (2.29) and ΔM_s refers to the quantum corrections to the mass. Using Eq. (2.116) we find the result

$$\Delta M_s = -\int_{-\infty}^{+\infty} d\rho\, \delta_* \mathcal{L}_{\text{eff}}[\varphi_S(\rho)] = \int_{-\infty}^{+\infty} \frac{dk}{4\pi} \sqrt{\alpha_0 + k^2}\, \frac{d\delta(k)}{dk}, \tag{2.124}$$

where we have used

$$\begin{aligned}\int_{-\infty}^{+\infty} d\rho\, U[\varphi_S(\rho)] &= \frac{2\alpha_0}{\beta^2}\int_{-\infty}^{+\infty} d\rho\, \frac{1}{\cosh^2(\sqrt{\alpha_0}\rho)} \\ &= \frac{2\sqrt{\alpha_0}}{\beta^2}\tanh(\sqrt{\alpha_0}\rho)\Big|_{-\infty}^{+\infty}\end{aligned}$$

$$= \frac{4\sqrt{\alpha_0}}{\beta^2}. \tag{2.125}$$

We want to remark that a finite contribution $-\sqrt{\alpha_0}/\pi$ (see Refs. [51, 79]) does not appear in Eq. (2.124) but we will discuss the appearance of such a finite contribution in a separate section below.

If we use the Lorentz covariant expression Eq. (2.117) we find for the correction to the soliton mass

$$\Delta M_s = -\int_{-\infty}^{+\infty} d\rho\, \delta_* \mathcal{L}_{\text{eff}}[\varphi_S(\rho)]$$
$$= -\int_{-\infty}^{+\infty} \frac{dk}{2\pi} \int_{-\infty}^{+\infty} \frac{d\omega}{2\pi i} \frac{2\sqrt{\alpha_0}}{\alpha_0 - \omega^2 + k^2 - i0}$$
$$= -\frac{\sqrt{\alpha_0}}{2\pi} \ln\left(\frac{\Lambda^2}{\alpha_0}\right), \tag{2.126}$$

where we have again used Eq. (2.125). Thus we find for the quantum mass of a soliton

$$M_s = \frac{8\sqrt{\alpha_0}}{\beta^2} - \frac{\sqrt{\alpha_0}}{2\pi} \ln\left(\frac{\Lambda^2}{\alpha_0}\right). \tag{2.127}$$

Renormalisation is performed by replacing α_0 by αZ_1, where Z_1 is defined by Eq. (2.44). So we find

$$M_s = \frac{8\sqrt{\alpha Z_1}}{\beta^2} - \frac{\sqrt{\alpha Z_1}}{2\pi} \ln\left(\frac{\Lambda^2}{\alpha Z_1}\right). \tag{2.128}$$

Using the expansion Eq. (2.120) and dropping all terms of order $O(\beta^2)$ in the soliton mass yields

$$M_s = \frac{8\sqrt{\alpha}}{\beta^2}\left[1 + \frac{\beta^2}{16\pi} \ln\left(\frac{\Lambda^2}{M^2}\right)\right] - \frac{\sqrt{\alpha}}{2\pi} \ln\left(\frac{\Lambda^2}{\alpha}\right)$$
$$= \frac{8\sqrt{\alpha}}{\beta^2} + \frac{\sqrt{\alpha}}{2\pi} \ln\left(\frac{\alpha}{M^2}\right). \tag{2.129}$$

In terms of the physical coupling constant Eq. (2.122) the soliton mass reads

$$M_s = \frac{8\sqrt{\alpha_{ph}}}{\beta^2}. \tag{2.130}$$

So we conclude that the soliton is renormalisable, does not depend on the normalisation scale M and is indeed an observable quantity. This corroborates the assertion by Zamolodchikov and Zamolodchikov [75], that the singularity of the sine–Gordon model due to a finite correction $-\sqrt{\alpha_{ph}}/\pi$ to the soliton mass, caused by Gaussian fluctuations around a soliton solution, is the result of a regularisation and renormalisation procedure.

2.8. Renormalisation of the soliton mass by Gaussian quantum corrections in discretised space–time

To corroborate the results obtained in continuous space–times we investigate the quantum corrections to the mass of a soliton by using a space–time discretisation approach (see [51, 79]). In this technique the system is put in a large box of spatial length L and length T in time direction with different boundary conditions as a constraint. The total number of discretised modes and the space–time volume of the box are sent to infinity at the end of the calculation thus approaching the continuum. The advantage of this technique is that the subtraction of the modes in the presence of the soliton and of the modes in trivial vacuum can be performed very accurately. Since this subtraction is rather subtle we have performed all calculations for periodic-, anti-periodic boundary conditions and for rigid walls. As can be seen the results agree in all cases perfectly well with the result obtained in continuous space–time.

Next we will separate the complex solutions of the quantum fluctuations stated in Eq. (2.105) into real and imaginary spatial parts. Since the time-dependent part does not affect the counting of the modes we set $\tau = 0$ in the explicit expressions below. In the soliton sector the real spatial part (this includes the zero mode) of Eq. (2.105) is symmetric under the transformation $\rho \leftrightarrow -\rho$ and will be denoted with an index "S". The imaginary part is anti-symmetric and is therefore denoted with "A"

$$\begin{aligned}
\phi_{k,S}^{(S)}(\rho) &= \tilde{C}_{k,S}^{(S)}(L)\,\mathfrak{Re}\,\phi_{\omega k}(0,\rho) \\
&= C_{k,S}^{(S)}(L)\left(k\cos(k\rho) - \sqrt{\alpha_0}\tanh(\sqrt{\alpha_0}\rho)\sin(k\rho)\right), \quad (2.131) \\
\phi_{b,S}^{(S)}(\rho) &= \tilde{C}_{b,S}^{(S)}(L)\,\mathfrak{Re}\,\phi_{\omega b}(0,\rho) \\
&= C_{b,S}^{(S)}(L)\frac{1}{\cosh(\sqrt{\alpha_0}\rho)}, \quad (2.132) \\
\phi_{k,A}^{(S)}(\rho) &= \tilde{C}_{k,A}^{(S)}(L)\,\mathfrak{Im}\,\phi_{\omega k}(0,\rho) \\
&= C_{k,A}^{(S)}(L)\left(k\sin(k\rho) + \sqrt{\alpha_0}\tanh(\sqrt{\alpha_0}\rho)\cos(k\rho)\right), \quad (2.133)
\end{aligned}$$

where the $\tilde{C}$ and C are functions of k and L such that all solutions are normalised to one for all values of the momenta and spatial distances between the boundaries. Furthermore, the time dependent part is assumed to be normalised to one for all values of ω and T. Since we do not need the explicit expressions of these constants and of the time dependent part we do not write them explicitly. The solution Eq. (2.132) corresponds to the discrete bound state. The asymptotic behaviour of the scattering solutions is given by

$$\begin{aligned}
\lim_{\rho\to\pm\infty}\phi_{k,S}^{(S)}(\rho) &= C_{k,S}^{(S)}(L)\cos\left(k\rho \pm \frac{1}{2}\delta(k)\right), \\
\lim_{\rho\to\pm\infty}\phi_{k,A}^{(S)}(\rho) &= C_{k,A}^{(S)}(L)\sin\left(k\rho \pm \frac{1}{2}\delta(k)\right),
\end{aligned} \quad (2.134)$$

with the phase shift

$$\delta(k) = 2\arctan\left(\frac{\sqrt{\alpha_0}}{k}\right). \tag{2.135}$$

The asymptotic limit for small momentum of $\delta(k)$ is given by

$$\delta(k) = \pi - \frac{2k}{\sqrt{\alpha_0}} + \mathcal{O}(k^3), \tag{2.136}$$

while the large momentum limit reads

$$\delta(k) = \frac{2\sqrt{\alpha_0}}{k} + \mathcal{O}\left(\frac{1}{k^3}\right). \tag{2.137}$$

In the vacuum sector we proceed analogous. After splitting the vacuum solution Eq. (2.111) in real and imaginary part we obtain the symmetric and antisymmetric solutions with respect to the transformations $\rho \leftrightarrow -\rho$

$$\begin{aligned}
\phi_{q,S}^{(V)}(\rho) &= \tilde{C}_{q,S}^{(V)}(L)\,\mathfrak{Re}\,\phi_{\omega q}(0,\rho) \\
&= C_{q,S}^{(V)}(L)\cos(q\rho), \tag{2.138} \\
\phi_{q,A}^{(V)}(\rho) &= \tilde{C}_{q,A}^{(V)}(L)\,\mathfrak{Im}\,\phi_{\omega q}(0,\rho) \\
&= C_{q,A}^{(V)}(L)\sin(k\rho), \tag{2.139}
\end{aligned}$$

where again $\tilde{C}$ and C normalise these functions to one. Since the boundary conditions in the case of all scattering solutions act only as a filter which selects out a discrete number of the continuum of modes, the eigenvalues and eigenfunctions of these modes are exactly the same as the corresponding modes in the continuum case, so to say the functions do not "feel" the boundaries. Contrary to this the bound state does indeed "feel" the boundaries which force some behaviour at $\pm L/2$ on the function, therefore the solution and the eigenvalue is slightly different for the different boundary conditions but in the limit $L \to \infty$ all cases coincide and approach the continuum values. In the following figures the continuum functions with the corresponding discretised momenta are depicted, this means that the scattering modes are represented exactly while the "real" bound state function looks slightly different for finite L.

We want to stress the well-known fact that each mode in the soliton sector has one corresponding mode in the vacuum and that it is crucial to subtract the modes carefully. Only this way it is guaranteed that the mass correction yields sensible results and that the theory is renormalisable. Next we calculate the determinant for three different boundary conditions and show that one obtains the same result as in the continuum limit.

2.8.1. Periodic boundary conditions

The periodic boundary conditions are given by

$$\phi(-L/2) = \phi(L/2),$$
$$\partial_\rho \phi(-L/2) = \partial_\rho \phi(L/2). \tag{2.140}$$

Substituting the corresponding solutions Eqs. (2.134), (2.138) in Eq. (2.140) the following conditions on the spatial momenta are obtained for the scattering solutions

$$k_n L + \delta(k_n) = 2\pi n \quad \text{in the soliton sector}, \tag{2.141}$$
$$q_n L = 2\pi n \quad \text{in the vacuum sector}, \tag{2.142}$$

where $n \in \mathbb{Z}$. Due to the complete (anti-)symmetry of all real solutions according to the transformation $k \to -k$ the pair $(k, -k)$ belongs to the same eigenfunction. Therefore, one has to count only either positive values of k and $k = 0$ or only negative values and $k = 0$ for the evaluation of the determinant. We choose the former alternative restricting the phase shift to $\delta(k) \in [0, \pi]$. It should be clear from Eqs. (2.141) and (2.142) that n is the number of nodes in the interval $[-L/2, L/2]$, while k_n is the spatial momentum of the fluctuation at $\rho \to \pm\infty$. Since the phase shift $\delta(k)$ has the same sign as k, k_n is lower than the corresponding vacuum moment q_n since the attractive potential effectively yields an extra spatial momentum $\delta(k_n)/L$.

Substituting Eq. (2.136) into Eq. (2.141) yields after an expansion in L with terms of order k_n^3 neglected the following result for the lowest momenta

$$k_n = \frac{(2n-1)\pi}{L}\left(1 + \frac{2}{L\sqrt{\alpha_0}}\right) + \mathcal{O}(L^{-3}). \tag{2.143}$$

There is no $n = 0$ scattering mode which is intuitively clear since this corresponds effectively to zero momentum. Due to the attraction the asymptotic momentum k_n should be negative to add up to zero, and indeed it is. The place of the $n = 0$ mode is taken by the bound state Eq. (2.132) which could be interpreted as a "trapped scattering state since it has asymptotic momentum $k_0 = 0$" (see Fig. 2.7 left) and yields the first logarithm in Eq. (2.148). It is also clear that in the case of periodic boundary conditions the bound state has no nodes. Since continuation to the interval $[L/2, 3L/2]$ yields an identical copy of the interval $[-L/2, L/2]$, the eigenfunction approaches but never crosses the ρ-axis of Fig. 2.7 left, thus yielding no nodes.

In the vacuum sector one can read off the momenta from Eq. (2.142)

$$q_n = \frac{2n\pi}{L}. \tag{2.144}$$

For $n = 0$ we have $q_0 = 0$. There is only one $n = 0$ mode, since the antisymmetric solution Eq. (2.139) vanishes for $q = 0$. The symmetric solution Eq. (2.138) yields the second logarithm in Eq. (2.148) and is depicted in Fig. 2.7 right.

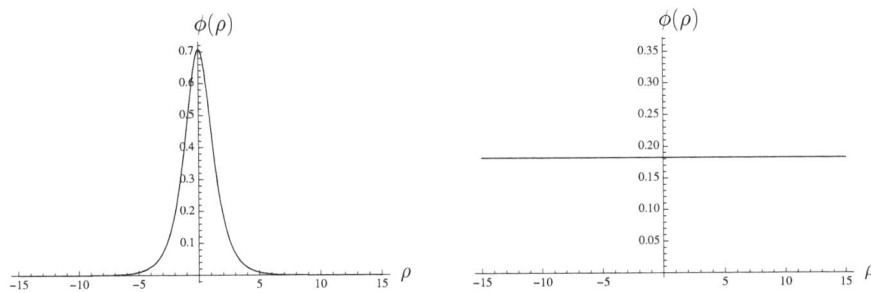

Figure 2.7.: To the *left* the zero mode with values $L = 30$ and $\alpha_0 = 1$ is depicted. To the *right* one can see the first ($n = 0$) mode in the vacuum with the same values for L and α_0.

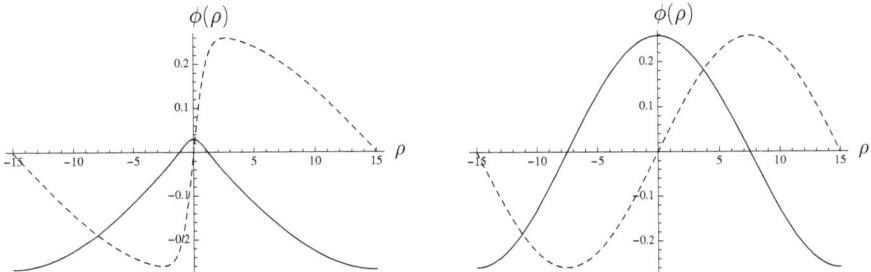

Figure 2.8.: Here the $n = 1$ modes are depicted. One can see the two modes in the soliton sector to the *left* and the corresponding vacuum modes to the *right* with values $L = 30$ and $\alpha_0 = 1$.

All higher modes ($n \geq 1$) appear doubly in the soliton sector and in the vacuum sector, once for the symmetric solution Eq. (2.131) and once for the antisymmetric solution Eq. (2.133). The $n = 1$ modes in both sectors are depicted in Fig. 2.8. The first momenta for all boundary conditions are given in Table 2.1.

We want to remark that the continuum limit of ω, respectively $M \to \infty$ and $T \to \infty$ can be performed trivially which is obvious since the presence of the soliton only affects the spatial momenta of the quantum fluctuations but not their energy. Therefore we consider only periodic boundary conditions in time direction. From the equations below it can be read off trivially that other boundary conditions in time direction yield the same result. Concerning the counting of modes we have to remark that if we consider the time dependent factor of all solutions proportional to the complex function $\exp(-i\omega\tau)$ we have to count all modes, i.e. the positive ones, the negative ones and the $\omega = 0$ mode since from $\int d\tau \exp(-i[\omega - \omega]\tau) = 0$ we see that the positive ω and negative $-\omega$ mode are orthogonal, i.e. belong to different eigenfunctions. If we split the time factor into real and imaginary part as we did for the spatial part we would have for each value of ω two functions, one proportional to $\sin(\omega\tau)$ and one proportional to $\cos(\omega\tau)$. But the $-\omega$ mode belongs now to the same eigenfunction as the ω mode and we have to count only positive values and $\omega = 0$ or negative values and $\omega = 0$. We choose the former variant for the complex time dependent factor hence taking positive and negative values of ω into account.

For periodic boundary conditions we find from $\exp(-i\omega\tau)$

$$\omega_m T = 2\pi m, \tag{2.145}$$

where $m \in \mathbb{Z}$. Furthermore, we want to stress that it is crucial to perform the limit $M \to \infty$ first and $T \to \infty$ only afterwards, respectively the limit $N \to \infty$ before $L \to \infty$, since otherwise the corresponding integration region vanishes.

Due to the relation Eq. (2.98)

$$\sqrt{\frac{\det(\Box + \alpha_0)}{\det(\Box + \alpha_0 \cos\beta\theta_{cl}(x))}} = \exp\left\{-\frac{1}{2}\sum_n \left(\ln \lambda_n^{(S)} - \ln \lambda_n^{(V)}\right)\right\}$$
$$= \exp(-i\Delta M_S T), \tag{2.146}$$

where the sums are purely discrete before the continuum limit, we find

$$\Delta M_S = \frac{1}{2iT} \sum_n \left(\ln \lambda_n^{(S)} - \ln \lambda_n^{(V)}\right), \tag{2.147}$$

respectively for the continuum limit

$$\Delta M_S = \lim_{M,T,N,L \to \infty} \frac{1}{2iT} \sum_{m=-M}^{M} \left[\ln(-\omega_m^2 - i0) - \ln(\alpha_0 - \omega_m^2 + q_0^2 - i0)\right.$$
$$\left. + 2\sum_{n=1}^{N} \ln(\alpha_0 - \omega_m^2 + k_n^2 - i0) - 2\sum_{n=1}^{N} \ln(\alpha_0 - \omega_m^2 + q_n^2 - i0)\right]. \tag{2.148}$$

Next we have to evaluate the sums. Using Eqs. (2.141), (2.142) and (2.145) we find

$$\sum_{n=1}^{N} \longrightarrow \int_{k_1}^{k_N} dk \frac{dn(k)}{dk} = \int_{\pi/L}^{(2\pi N - \delta)/L} \frac{dk}{2\pi}\left(L + \frac{d\delta(k)}{dk}\right), \tag{2.149}$$

$$\sum_{m=-M}^{M} \longrightarrow \int_{-\omega_M}^{\omega_M} d\omega \frac{dm(\omega)}{d\omega} = T\int_{-2\pi M/T}^{2\pi M/T} \frac{d\omega}{2\pi}, \tag{2.150}$$

in the soliton sector, where we have dropped terms of order $\mathcal{O}(L^{-2})$, and

$$\sum_{n=1}^{N} \longrightarrow \int_{q_1}^{q_N} dq \frac{dn(q)}{dq} = \int_{2\pi/L}^{2\pi N/L} \frac{dq}{2\pi} L, \tag{2.151}$$

$$\sum_{m=-M}^{M} \longrightarrow T\int_{-2\pi M/T}^{2\pi M/T} \frac{d\omega}{2\pi}, \tag{2.152}$$

in the vacuum sector. We want to remark that the integrals on the right are obtained only in the limit $L \to \infty$. For finite L the integral representation is only approximate. Substituting the integral limits of these sums in Eq. (2.148) we find

$$\Delta M_S = \lim_{N,L\to\infty} \frac{1}{2i} \int_{-\infty}^{\infty} \frac{d\omega}{2\pi} \bigg[\ln(-\omega^2 - i0) - \ln(\alpha_0 - \omega^2 - i0)$$
$$+ 2\int_{\pi/L}^{(2\pi N-\delta)/L} \frac{dk}{2\pi} \left(L + \frac{d\delta(k)}{dk}\right) \ln(\alpha_0 - \omega^2 + k^2 - i0)$$
$$- 2\int_{2\pi/L}^{2\pi N/L} \frac{dq}{2\pi} L \ln(\alpha_0 - \omega^2 + q^2 - i0) \bigg], \qquad (2.153)$$

where we have used that $q_0^2 = 0$ in the first line, respectively

$$\Delta M_S = \lim_{N,L\to\infty} \frac{1}{2i} \int_{-\infty}^{\infty} \frac{d\omega}{2\pi} \bigg[\ln(-\omega^2 - i0) - \ln(\alpha_0 - \omega^2 - i0)$$
$$+ 2\int_{\pi/L}^{2\pi/L} \frac{dk}{2\pi} \left(L + \frac{d\delta(k)}{dk}\right) \ln(\alpha_0 - \omega^2 + k^2 - i0)$$
$$+ 2\int_{2\pi/L}^{(2\pi N-\delta)/L} \frac{dk}{2\pi} \frac{d\delta(k)}{dk} \ln(\alpha_0 - \omega^2 + k^2 - i0)$$
$$- 2\int_{(2\pi N-\delta)/L}^{2\pi N/L} \frac{dk}{2\pi} L \ln(\alpha_0 - \omega^2 + k^2 - i0) \bigg]. \qquad (2.154)$$

We will perform the limits of the last three logarithms above individually

$$\lim_{N,L\to\infty} 2\int_{\pi/L}^{2\pi/L} \frac{dk}{2\pi} \left(L + \frac{d\delta(k)}{dk}\right) \ln(\alpha_0 - \omega^2 + k^2 - i0) =$$
$$= \lim_{N,L\to\infty} 2\frac{\pi}{L}\frac{L}{2\pi} \ln\left(\alpha_0 - \omega^2 + \left(\frac{2\pi}{L} - i0\right)^2\right)$$
$$= \ln(\alpha_0 - \omega^2 - i0), \qquad (2.155)$$

where we have set $k = 2\pi/L$ the upper bound value. Obviously, the result is the same for any value of k in the interval $[\pi/L, 2\pi/L]$. Next we have

$$\lim_{N,L\to\infty} 2\int_{2\pi/L}^{(2\pi N-\delta)/L} \frac{dk}{2\pi} \frac{d\delta(k)}{dk} \ln(\alpha_0 - \omega^2 + k^2 - i0) =$$
$$= 2\int_0^{\infty} \frac{dk}{2\pi} \frac{d\delta(k)}{dk} \ln(\alpha_0 - \omega^2 + k^2 - i0), \qquad (2.156)$$

and finally

$$\lim_{N,L\to\infty} -2\int_{(2\pi N-\delta)/L}^{2\pi N/L} \frac{dk}{2\pi} L \ln(\alpha_0 - \omega^2 + k^2 - i0) =$$
$$= \lim_{N,L\to\infty} -2\frac{\sqrt{\alpha_0}}{\pi N}\frac{L}{2\pi} \ln\left(\alpha_0 - \omega^2 + \left(\frac{2\pi N}{L}\right)^2 - i0\right)$$
$$= 0, \qquad (2.157)$$

since the limit $N \to \infty$ has to be performed before the limit $L \to \infty$. Furthermore, we have used Eq. (2.137) to find $\delta = \sqrt{\alpha_0} L / \pi N$. Substituting these results in Eq. (2.154) yields

$$\Delta M_S = \frac{1}{2i} \int_{-\infty}^{\infty} \frac{d\omega}{2\pi} \left[\ln(-\omega^2 - i0) + 2 \int_0^{\infty} \frac{dk}{2\pi} \frac{d\delta(k)}{dk} \ln(\alpha_0 - \omega^2 + k^2 - i0) \right]. \qquad (2.158)$$

Multiplying the first logarithm, which is due to the bound state, with

$$1 = \int_{-\infty}^{\infty} \frac{dk}{2\pi} \frac{2\sqrt{\alpha_0}}{\alpha_0 + k^2}, \qquad (2.159)$$

and using

$$\frac{d\delta(k)}{dk} = -\frac{2\sqrt{\alpha_0}}{\alpha_0 + k^2}, \qquad (2.160)$$

the mass correction reads

$$\Delta M_S = \int_{-\infty}^{\infty} \frac{d\omega}{2\pi i} \int_{-\infty}^{\infty} \frac{dk}{2\pi} \frac{\sqrt{\alpha_0}}{\alpha_0 + k^2} \left(\ln(-\omega^2 - i0) - \ln(\alpha_0 - \omega^2 + k^2 - i0) \right). \qquad (2.161)$$

After partial integration with respect to ω we find

$$\Delta M_S = -\int_{-\infty}^{\infty} \frac{d\omega}{2\pi i} \int_{-\infty}^{\infty} \frac{dk}{2\pi} \frac{2\sqrt{\alpha_0}}{\alpha_0 - \omega^2 + k^2 - i0}. \qquad (2.162)$$

The result obtained for periodic boundary conditions equals the result obtained in continuous space–time Eq. (2.126). The evaluation of the mass correction using other boundary conditions runs parallel and will be performed next.

2.8.2. Anti-periodic boundary conditions

The anti-periodic boundary conditions are given by

$$\phi(-L/2) = -\phi(L/2), \\ \partial_\rho \phi(-L/2) = -\partial_\rho \phi(L/2). \qquad (2.163)$$

We obtain for the scattering solutions

$$k_n L + \delta(k_n) = (2n-1)\pi \quad \text{in the soliton sector}, \qquad (2.164)$$
$$q_n L = (2n-1)\pi \quad \text{in the vacuum sector}, \qquad (2.165)$$

where $n \in \mathbb{Z}$. For the momenta we find

$$k_n = \frac{2(n-1)\pi}{L} \left(1 + \frac{2}{L\sqrt{\alpha_0}} \right) + \mathcal{O}(L^{-3}). \qquad (2.166)$$

There is no $n = 0$ mode for anti-periodic boundary conditions neither in the soliton sector nor in the vacuum sector, which is intuitively clear since continuation to the interval

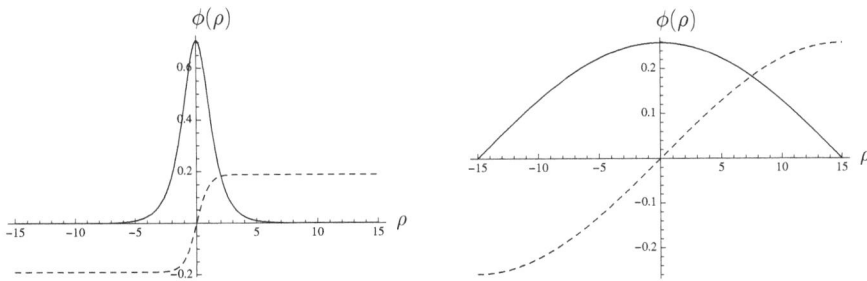

Figure 2.9.: To the *left* one can see the single non-vanishing $n = 1$ scattering mode in the soliton sector and the bound state with values $L = 30$ and $\alpha_0 = 1$. To the *right* one can see the first $n = 1$ mode pair in the vacuum sector with the same values for L and α_0.

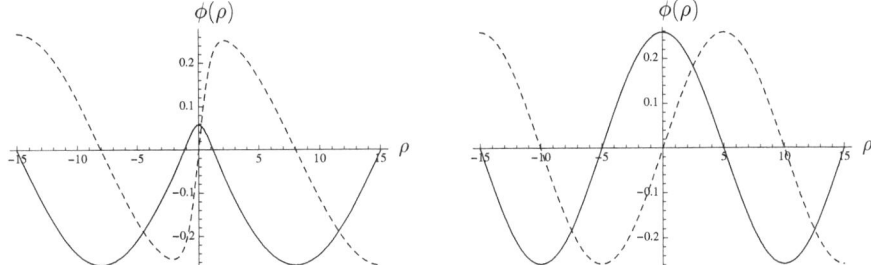

Figure 2.10.: Here the $n = 2$ modes are depicted. One can see the two modes in the soliton sector to the *left* and the corresponding vacuum modes to the *right* with values $L = 30$ and $\alpha_0 = 1$.

$[L/2, 3L/2]$ yields an identical copy of the interval $[-L/2, L/2]$ up to the transformation $\phi \to -\phi$. Therefore, all eigenfunctions have to cross the ρ-axis yielding non-zero nodes. The first modes are $n = 1$ modes with $k_1 = 0$. The bound state is a $n = 1$ mode since it crosses once the ρ-axis per interval $[\rho, \rho + L]$ and since it has "asymptotic momentum $k_1 = 0$". From the scattering solutions only the antisymmetric one Eq. (2.133) is non-vanishing for $n = 1$. Both $n = 1$ solutions are depicted in Fig. 2.9 *left* yielding the first two logarithms in Eq. (2.168). From $n \geq 2$ all modes appear doubly, once the symmetric solution and once the antisymmetric one. The two $n = 2$ modes are depicted in Fig. 2.10 *left*.

In the vacuum sector we find for the momenta

$$q_n = \frac{(2n-1)\pi}{L}. \qquad (2.167)$$

In the vacuum all modes appear doubly from $n \geq 1$. The two $n = 1$ modes are depicted in Fig. 2.9 *right* and yield the third logarithm in Eq. (2.168), while the two $n = 2$ modes are depicted in Fig. 2.10 *right*. The first values for the momenta in both sectors can be found in Table 2.1.

For anti-periodic boundary conditions we find for the quantum correction to the mass

$$\Delta M_S = \lim_{M,T,N,L\to\infty} \frac{1}{2iT} \sum_{m=-M}^{M} \left[\ln(-\omega_m^2 - i0) + \ln(\alpha_0 - \omega_m^2 + k_1^2 - i0) - 2\ln(\alpha_0 - \omega_m^2 + q_1^2 - i0) \right.$$
$$\left. + 2\sum_{n=2}^{N} \ln(\alpha_0 - \omega_m^2 + k_n^2 - i0) - 2\sum_{n=2}^{N} \ln(\alpha_0 - \omega_m^2 + q_n^2 - i0) \right]. \quad (2.168)$$

The sums read for anti-periodic boundary conditions

$$\sum_{n=2}^{N} \longrightarrow \int_{k_2}^{k_N} dk \frac{dn(k)}{dk} = \int_{2\pi/L}^{((2N-1)\pi-\delta)/L} \frac{dk}{2\pi} \left(L + \frac{d\delta(k)}{dk} \right), \quad (2.169)$$

$$\sum_{m=-M}^{M} \longrightarrow T \int_{-2\pi M/T}^{2\pi M/T} \frac{d\omega}{2\pi}, \quad (2.170)$$

in the soliton sector, where we have again dropped terms of order $\mathcal{O}(L^{-2})$, and

$$\sum_{n=2}^{N} \longrightarrow \int_{q_2}^{q_N} dq \frac{dn(q)}{dq} = \int_{3\pi/L}^{(2N-1)\pi/L} \frac{dq}{2\pi} L, \quad (2.171)$$

$$\sum_{m=-M}^{M} \longrightarrow T \int_{-2\pi M/T}^{2\pi M/T} \frac{d\omega}{2\pi}, \quad (2.172)$$

in the vacuum sector. For the mass correction we find

$$\Delta M_S = \lim_{N,L\to\infty} \frac{1}{2i} \int_{-\infty}^{\infty} \frac{d\omega}{2\pi} \left[\ln(-\omega^2 - i0) + \ln(\alpha_0 - \omega^2 - i0) \right.$$
$$- 2\ln(\alpha_0 - \omega^2 + q_1^2 - i0)$$
$$+ 2\int_{2\pi/L}^{((2N-1)\pi-\delta)/L} \frac{dk}{2\pi} \left(L + \frac{d\delta(k)}{dk} \right) \ln(\alpha_0 - \omega^2 + k^2 - i0)$$
$$\left. - 2\int_{3\pi/L}^{(2N-1)\pi/L} \frac{dq}{2\pi} L \ln(\alpha_0 - \omega^2 + q^2 - i0) \right], \quad (2.173)$$

where we have used that $k_1^2 = 0$ in the first line, respectively

$$\Delta M_S = \lim_{N,L\to\infty} \frac{1}{2i} \int_{-\infty}^{\infty} \frac{d\omega}{2\pi} \left[\ln(-\omega^2 - i0) + \ln(\alpha_0 - \omega^2 - i0) \right.$$
$$- 2\ln(\alpha_0 - \omega^2 + q_1^2 - i0)$$
$$+ 2\int_{2\pi/L}^{3\pi/L} \frac{dk}{2\pi} \left(L + \frac{d\delta(k)}{dk} \right) \ln(\alpha_0 - \omega^2 + k^2 - i0)$$
$$+ 2\int_{3\pi/L}^{((2N-1)\pi-\delta)/L} \frac{dk}{2\pi} \frac{d\delta(k)}{dk} \ln(\alpha_0 - \omega^2 + k^2 - i0)$$
$$\left. - 2\int_{((2N-1)\pi-\delta)/L}^{(2N-1)\pi/L} \frac{dk}{2\pi} L \ln(\alpha_0 - \omega^2 + k^2 - i0) \right]. \quad (2.174)$$

The last three logarithms yield the same result as the corresponding expressions for periodic boundary conditions

$$\lim_{N,L\to\infty} 2\int_{2\pi/L}^{3\pi/L} \frac{dk}{2\pi}\left(L + \frac{d\delta(k)}{dk}\right)\ln(\alpha_0 - \omega^2 + k^2 - i0) = \ln(\alpha_0 - \omega^2 - i0), \quad (2.175)$$

$$\lim_{N,L\to\infty} 2\int_{3\pi/L}^{((2N-1)\pi-\delta)/L} \frac{dk}{2\pi}\frac{d\delta(k)}{dk}\ln(\alpha_0 - \omega^2 + k^2 - i0) = 2\int_0^\infty \frac{dk}{2\pi}\frac{d\delta(k)}{dk}\ln(\alpha_0 - \omega^2 + k^2 - i0), \quad (2.176)$$

$$\lim_{N,L\to\infty} -2\int_{((2N-1)\pi-\delta)/L}^{(2N-1)\pi/L} \frac{dk}{2\pi} L\ln(\alpha_0 - \omega^2 + k^2 - i0) = 0. \quad (2.177)$$

Since $\lim_{N,L\to\infty}\ln(\alpha_0 - \omega^2 + q_1^2 - i0) = \ln(\alpha_0 - \omega^2 - i0)$ we find again

$$\Delta M_S = \frac{1}{2i}\int_{-\infty}^\infty \frac{d\omega}{2\pi}\left[\ln(-\omega^2 - i0) + 2\int_0^\infty \frac{dk}{2\pi}\frac{d\delta(k)}{dk}\ln(\alpha_0 - \omega^2 + k^2 - i0)\right], \quad (2.178)$$

and thus

$$\Delta M_S = -\int_{-\infty}^\infty \frac{d\omega}{2\pi i}\int_{-\infty}^\infty \frac{dk}{2\pi}\frac{2\sqrt{\alpha_0}}{\alpha_0 - \omega^2 + k^2 - i0}. \quad (2.179)$$

The result obtained for anti-periodic boundary conditions equals the result obtained in continuous space–time Eq. (2.126) and the discretised result for periodic boundary conditions Eq. (2.162).

2.8.3. Rigid walls

Rigid walls are given by

$$\phi(-L/2) = \phi(L/2) = 0. \quad (2.180)$$

For the scattering solutions we find

$$k_n L + \delta(k_n) = n\pi \quad \text{in the soliton sector}, \quad (2.181)$$
$$q_n L = n\pi \quad \text{in the vacuum sector}, \quad (2.182)$$

where $n \in \mathbb{Z}$. For rigid walls the first spatial momenta in the soliton sector read

$$k_n = \frac{(n-1)\pi}{L}\left(1 + \frac{2}{L\sqrt{\alpha_0}}\right) + \mathcal{O}(L^{-3}). \quad (2.183)$$

There is no $n = 0$ mode in the soliton sector since the corresponding momentum k_0 is negative. For $n = 1$ ($k_1 = 0$) there are no scattering solutions since the symmetric mode vanishes and the antisymmetric one does not agree with the boundary conditions. They are replaced by the bound state which is a $n = 1$ mode since $k_1 = 0$ and yields the first logarithm in Eq. (2.185) (see Fig. 2.11 *left*).

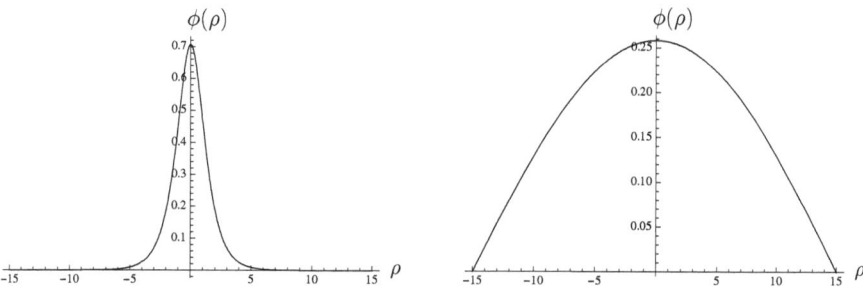

Figure 2.11.: To the *left* the bound state is depicted with values $L = 30$ and $\alpha_0 = 1$ which is the only $n = 1$ mode in the soliton sector. To the *right* the single $n = 1$ vacuum mode with the same values can be seen.

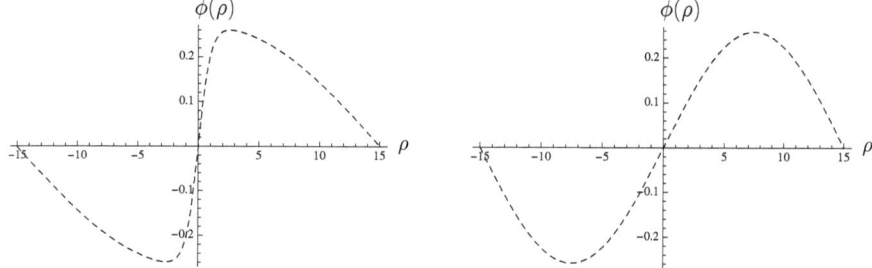

Figure 2.12.: *Left* the $n = 2$ soliton mode with values $L = 30$ and $\alpha_0 = 1$ and *right* the corresponding vacuum mode with the same values are depicted.

For $n \geq 2$ each scattering mode appears only single, since alternating in n only the symmetric or antisymmetric solution agrees with the boundary conditions. The $n = 2$ solution which is due to Eq. (2.133) is depicted in Fig. 2.12 *left*.

In the vacuum sector we find

$$q_n = \frac{n\pi}{L}. \tag{2.184}$$

Obviously, there is no $n = 0$ mode since the antisymmetric mode vanishes and the symmetric one does not agree with the boundary conditions. For $n \geq 1$ each scattering mode appears only single by the same reason as in the soliton sector. The first mode is the $n = 1$ solution which is due to Eq. (2.138) and yields the second logarithm in Eq. (2.185) (see Fig. 2.11 *right*). The single $n = 2$ due to Eq. (2.139) is depicted in Fig. 2.12 *right*. The first momenta can be found in Table 2.1.

For rigid walls the quantum correction to the mass reads

$$\Delta M_S = \lim_{M,T,N,L \to \infty} \frac{1}{2iT} \sum_{m=-M}^{M} \left[\ln(-\omega_m^2 - i0) - \ln(\alpha_0 - \omega_m^2 + q_1^2 - i0) \right.$$

$$+ \sum_{n=2}^{N} \ln(\alpha_0 - \omega_m^2 + k_n^2 - i0) - \sum_{n=2}^{N} \ln(\alpha_0 - \omega_m^2 + q_n^2 - i0) \bigg]. \tag{2.185}$$

For the sums we find

$$\sum_{n=2}^{N} \longrightarrow \int_{k_2}^{k_N} dk \, \frac{dn(k)}{dk} = \int_{\pi/L}^{(N\pi-\delta)/L} \frac{dk}{\pi} \left(L + \frac{d\delta(k)}{dk} \right), \tag{2.186}$$

$$\sum_{m=-M}^{M} \longrightarrow T \int_{-2\pi M/T}^{2\pi M/T} \frac{d\omega}{2\pi}, \tag{2.187}$$

in the soliton sector, where we have again dropped terms of order $\mathcal{O}(L^{-2})$. In the vacuum sector we find

$$\sum_{n=2}^{N} \longrightarrow \int_{q_2}^{q_N} dq \, \frac{dn(q)}{dq} = \int_{2\pi/L}^{N\pi/L} \frac{dq}{\pi} L, \tag{2.188}$$

$$\sum_{m=-M}^{M} \longrightarrow T \int_{-2\pi M/T}^{2\pi M/T} \frac{d\omega}{2\pi}. \tag{2.189}$$

The mass correction reads

$$\Delta M_S = \lim_{N,L \to \infty} \frac{1}{2i} \int_{-\infty}^{\infty} \frac{d\omega}{2\pi} \bigg[\ln(-\omega^2 - i0) - \ln(\alpha_0 - \omega^2 + q_1^2 - i0) \\ + \int_{\pi/L}^{(N\pi-\delta)/L} \frac{dk}{\pi} \left(L + \frac{d\delta(k)}{dk} \right) \ln(\alpha_0 - \omega^2 + k^2 - i0) \\ - \int_{2\pi/L}^{N\pi/L} \frac{dq}{\pi} L \ln(\alpha_0 - \omega^2 + q^2 - i0) \bigg], \tag{2.190}$$

respectively

$$\Delta M_S = \lim_{N,L \to \infty} \frac{1}{2i} \int_{-\infty}^{\infty} \frac{d\omega}{2\pi} \bigg[\ln(-\omega^2 - i0) - \ln(\alpha_0 - \omega^2 + q_1^2 - i0) \\ + \int_{\pi/L}^{2\pi/L} \frac{dk}{\pi} \left(L + \frac{d\delta(k)}{dk} \right) \ln(\alpha_0 - \omega^2 + k^2 - i0) \\ + \int_{2\pi/L}^{(N\pi-\delta)/L} \frac{dk}{\pi} \frac{d\delta(k)}{dk} \ln(\alpha_0 - \omega^2 + k^2 - i0) \\ - \int_{(N\pi-\delta)/L}^{N\pi/L} \frac{dk}{\pi} L \ln(\alpha_0 - \omega^2 + k^2 - i0) \bigg]. \tag{2.191}$$

The last three logarithms yield the same result as the corresponding expressions for periodic and anti-periodic boundary conditions

$$\lim_{N,L \to \infty} \int_{\pi/L}^{2\pi/L} \frac{dk}{\pi} \left(L + \frac{d\delta(k)}{dk} \right) \ln(\alpha_0 - \omega^2 + k^2 - i0) = \ln(\alpha_0 - \omega^2 - i0), \tag{2.192}$$

$$\lim_{N,L \to \infty} \int_{2\pi/L}^{(N\pi-\delta)/L} \frac{dk}{\pi} \frac{d\delta(k)}{dk} \ln(\alpha_0 - \omega^2 + k^2 - i0) = \int_{0}^{\infty} \frac{dk}{\pi} \frac{d\delta(k)}{dk} \ln(\alpha_0 - \omega^2 + k^2 - i0),$$

(2.193)
$$\lim_{N,L\to\infty} -\int_{(N\pi-\delta)/L}^{N\pi/L} \frac{dk}{\pi} L \ln(\alpha_0 - \omega^2 + k^2 - i0) = 0\,. \tag{2.194}$$

Using $\lim_{N,L\to\infty} \ln(\alpha_0 - \omega^2 + q_1^2 - i0) = \ln(\alpha_0 - \omega^2 - i0)$ we find

$$\Delta M_S = \frac{1}{2i} \int_{-\infty}^{\infty} \frac{d\omega}{2\pi} \left[\ln(-\omega^2 - i0) + \int_0^{\infty} \frac{dk}{\pi} \frac{d\delta(k)}{dk} \ln(\alpha_0 - \omega^2 + k^2 - i0) \right], \tag{2.195}$$

respectively

$$\Delta M_S = -\int_{-\infty}^{\infty} \frac{d\omega}{2\pi i} \int_{-\infty}^{\infty} \frac{dk}{2\pi} \frac{2\sqrt{\alpha_0}}{\alpha_0 - \omega^2 + k^2 - i0}\,. \tag{2.196}$$

This result equals the results obtained for periodic boundary conditions Eq. (2.162) and anti-periodic boundary conditions Eq. (2.179) and corroborates the result obtained in continuous space–time Eq. (2.126).

2.8.4. Remark on the additional finite correction to the soliton mass

Finally, we want to remark how one can obtain the additional finite contribution $-\sqrt{\alpha_0}/\pi$ to the mass, which appears in the literature.

According to Ref. [80] the following relation holds

$$\sum_{m=-\infty}^{+\infty} \left[\ln(-\omega_m^2 + a^2) - \ln(-\omega_m^2 + b^2) \right] = i(a-b)T + 2\ln\left(\frac{1-e^{-iaT}}{1-e^{-ibT}}\right). \tag{2.197}$$

We perform the calculation only for periodic boundary conditions. The other boundary conditions yield the same result.

We start again with Eq. (2.148) but make use of Eq. (2.197)

$$\Delta M_S = \lim_{M,T,N,L\to\infty} \frac{1}{2iT} \sum_{m=-M}^{M} \Big[\ln(-\omega_m^2 - i0) - \ln(\alpha_0 - \omega_m^2 + q_0^2 - i0)$$
$$+ 2\sum_{n=1}^{N} \ln(\alpha_0 - \omega_m^2 + k_n^2 - i0) - 2\sum_{n=1}^{N} \ln(\alpha_0 - \omega_m^2 + q_n^2 - i0) \Big]$$
$$= -\frac{\sqrt{\alpha_0}}{2} + \lim_{N,L\to\infty} \sum_{n=1}^{N} \left(\sqrt{\alpha_0 + k_n^2} - \sqrt{\alpha_0 + q_n^2} \right), \tag{2.198}$$

where we did not denote $-i0$ in the last line. For the sums we have

$$\sum_{n=1}^{N} \longrightarrow \int_{k_1}^{k_N} dk \frac{dn(k)}{dk} = \int_{\pi/L}^{(2\pi N-\delta)/L} \frac{dk}{2\pi} \left(L + \frac{d\delta(k)}{dk} \right), \tag{2.199}$$

$$\sum_{n=1}^{N} \longrightarrow \int_{q_1}^{q_N} dq \frac{dn(q)}{dq} = \int_{2\pi/L}^{2\pi N/L} \frac{dq}{2\pi} L\,, \tag{2.200}$$

where we have dropped terms of order $\mathcal{O}(L^{-2})$. Substituting the integral limits of these

Renormalisation of the soliton mass in discretised space–time   55

sums we find

$$\Delta M_S = -\frac{\sqrt{\alpha_0}}{2} + \lim_{N,L\to\infty} \int_{\pi/L}^{(2\pi N-\delta)/L} \frac{dk}{2\pi}\left(L + \frac{d\delta(k)}{dk}\right)\sqrt{\alpha_0 + k^2} - \lim_{N,L\to\infty} \int_{2\pi/L}^{2\pi N/L} \frac{dq}{2\pi} L\sqrt{\alpha_0 + q^2}$$

$$= -\frac{\sqrt{\alpha_0}}{2} + \lim_{N,L\to\infty} \int_{\pi/L}^{2\pi/L} \frac{dk}{2\pi}\left(L + \frac{d\delta(k)}{dk}\right)\sqrt{\alpha_0 + k^2} + \lim_{N,L\to\infty} \int_{2\pi/L}^{(2\pi N-\delta)/L} \frac{dk}{2\pi}\frac{d\delta(k)}{dk}\sqrt{\alpha_0 + k^2}$$

$$- \lim_{N,L\to\infty} \int_{(2\pi N-\delta)/L}^{2\pi N/L} \frac{dk}{2\pi} L\sqrt{\alpha_0 + k^2}. \qquad (2.201)$$

Evaluating the last three integrals yields

$$\lim_{N,L\to\infty} \int_{\pi/L}^{2\pi/L} \frac{dk}{2\pi}\left(L + \frac{d\delta(k)}{dk}\right)\sqrt{\alpha_0 + k^2} = \lim_{N,L\to\infty} \frac{\pi}{L}\frac{L}{2\pi}\sqrt{\alpha_0 + \left(\frac{2\pi}{L}\right)^2}$$

$$= \frac{\sqrt{\alpha_0}}{2}, \qquad (2.202)$$

where we have set $k = 2\pi/L$ the upper bound value. Obviously, the result is the same for any value of k in the interval $[\pi/L, 2\pi/L]$. Next we find

$$\lim_{N,L\to\infty} \int_{2\pi/L}^{(2\pi N-\delta)/L} \frac{dk}{2\pi}\frac{d\delta(k)}{dk}\sqrt{\alpha_0 + k^2} = \int_0^\infty \frac{dk}{2\pi}\frac{d\delta(k)}{dk}\sqrt{\alpha_0 + k^2}, \qquad (2.203)$$

and finally

$$\lim_{N,L\to\infty} -\int_{(2\pi N-\delta)/L}^{2\pi N/L} \frac{dk}{2\pi} L\sqrt{\alpha_0 + k^2} = \lim_{N,L\to\infty} -\frac{\sqrt{\alpha_0}}{\pi N}\frac{L}{2\pi}\sqrt{\alpha_0 + \left(\frac{2\pi N}{L}\right)^2}$$

$$= -\frac{\sqrt{\alpha_0}}{\pi}, \qquad (2.204)$$

since the limit $N \to \infty$ has to be performed before the limit $L \to \infty$. Furthermore, we have used again Eq. (2.137) to find $\delta = \sqrt{\alpha_0}L/\pi N$. Substituting these results yields

$$\Delta M_S = \int_{-\infty}^\infty \frac{dk}{4\pi}\frac{d\delta(k)}{dk}\sqrt{\alpha_0 + k^2} - \frac{\sqrt{\alpha_0}}{\pi}. \qquad (2.205)$$

This result does not agree with the results obtained in continuous space–time Eq. (2.124) and in the discretised space–time approach Eqs. (2.162), (2.179) and (2.196) due to the additional finite contribution $-\sqrt{\alpha_0}/\pi$.

The finite contribution coming from Eq. (2.204) corresponds to preceding Eqs. (2.157), (2.177) and (2.194) in the discretised space–time approach performed before. One can see that in the original calculation of the preceding sections this finite contribution vanishes for the three boundary conditions. We want to remark that a non-vanishing finite contribution appears only if the sum over m, which corresponds to the integration over ω, is performed first and the continuum limit $M, T, N, L \to \infty$ is performed afterwards. If the continuum limit is performed first and the integration over ω is done afterwards this finite contribution vanishes.

In a strict sense the continuum limit, which matters is the limit $N, L \to \infty$, i.e. the continuum limit of the momentum k, which is intuitively clear from the discussion concerning the relevance of the discretisation of ω just before Eq. (2.145). This means that we get a non-vanishing finite contribution if we perform the sum over m first and the continuum limit $M, T, N, L \to \infty$ afterwards or if we perform the continuum limit $M, T \to \infty$ first, perform a residuum integration over ω next and perform the continuum limit $N, L \to \infty$ at last. On the contrary a vanishing finite contribution is obtained if the continuum limit $N, L \to \infty$ is carried out first.

In a Lorentz-invariant renormalisation procedure the frequency ω and spatial momentum k form a Lorentz-invariant 2-vector in Minkowski space–time. Therefore, it is of crucial importance to deal with ω and k on an equal footing at all steps of a calculation. This means that a Lorentz-invariant renormalisation procedure is conform only with performing the continuum limit $M, T, N, L \to \infty$ first, while the integration over ω is done afterwards, yielding a vanishing finite contribution. Thus we find that no finite correction to the quantum mass of as soliton appears in a Lorentz-invariant renormalisation procedure. This corroborates further our results obtained in continuous space–time and the fact that Gaussian quantum fluctuations do not change the mass of a soliton.

PERIODIC BC:

Soliton Sector		Vacuum Sector
$k_n L + \delta(k_n) = 2n\pi$		$q_n L = 2n\pi$
$n = 0 : (*)$	$\leftarrow 1\times \rightarrow$	$n = 0 : q_0 = 0$
$n = 1 : k_1 = \pi/L + \mathcal{O}(1/L^2)$	$\leftarrow 2\times \rightarrow$	$n = 1 : q_1 = 2\pi/L$
$n = 2 : k_2 = 3\pi/L + \mathcal{O}(1/L^2)$	$\leftarrow 2\times \rightarrow$	$n = 2 : q_2 = 4\pi/L$
...		...

ANTI-PERIODIC BC:

Soliton Sector		Vacuum Sector
$k_n L + \delta(k_n) = (2n-1)\pi$		$q_n L = (2n-1)\pi$
$n = 1 : (*) + k_1 = 0$	$\leftarrow (1+1)\times \rightarrow$	$n = 1 : q_1 = \pi/L$
$n = 2 : k_2 = 2\pi/L + \mathcal{O}(1/L^2)$	$\leftarrow 2\times \rightarrow$	$n = 2 : q_2 = 3\pi/L$
$n = 3 : k_3 = 4\pi/L + \mathcal{O}(1/L^2)$	$\leftarrow 2\times \rightarrow$	$n = 3 : q_3 = 5\pi/L$
...		...

RIGID WALLS:

Soliton Sector		Vacuum Sector
$k_n L + \delta(k_n) = n\pi$		$q_n L = n\pi$
$n = 1 : (*)$	$\leftarrow 1\times \rightarrow$	$n = 1 : q_1 = \pi/L$
$n = 2 : k_2 = \pi/L + \mathcal{O}(1/L^2)$	$\leftarrow 1\times \rightarrow$	$n = 2 : q_2 = 2\pi/L$
$n = 3 : k_3 = 2\pi/L + \mathcal{O}(1/L^2)$	$\leftarrow 1\times \rightarrow$	$n = 3 : q_3 = 3\pi/L$
...		...

Table 2.1.: The spectra of the momenta of the first quantum corrections of a soliton and the trivial vacuum. The modes, denoted by $(*)$ are due to the bound state.

2.9. Conclusion

The renormalisability of the sine–Gordon model was investigated by analysing the renormalisability of the two–point Green function to second order in α and to all orders in β^2. It was shown that the divergences appearing in the sine–Gordon model can be removed by the renormalisation of the dimensional coupling constant $\alpha_0(\Lambda^2)$. We have found that the coupling constant β^2 is not affected by quantum corrections which agrees well with the interpretation of β^2 as $\hbar$ (see Refs. [10, 30]). We have shown that all divergences can be absorbed by the coupling constant α depending on the normalisation scale M. Quantum fluctuations around the trivial vacuum calculated to first order in α and to all orders in β^2 form a physical coupling constant α_{ph} which is finite and independent of the normalisation scale M. We have shown that the total renormalised two–point Green function depends solely on the physical coupling constant α_{ph} by calculating the correction to the two–point Green function to second order in α and all orders in β^2.

Furthermore, we have analysed the renormalisability of the sine–Gordon model with respect to quantum fluctuations around a soliton. We have taken into account only Gaussian fluctuations (see Refs. [71, 72, 73]).

For the calculation of the effective Lagrangian, induced by Gaussian fluctuations, we have performed a split of the field in a classical field which is a solution of the equation of motion and a fluctuating field. The path–integral was performed over the fluctuating field to Gaussian order. It was shown that the renormalised effective Lagrangian, due to Gaussian fluctuations around a soliton, agrees completely with the renormalised Lagrangian of the sine–Gordon model, caused by quantum fluctuations around the trivial vacuum to first order in α and to second order in β^2. We have found that after the renormalisation the soliton mass is equal to the mass of a soliton, calculated without quantum corrections, up to the replacement $\alpha_0 \to \alpha_{ph}$. So we conclude that Gaussian fluctuations around a soliton do not produce quantum corrections to the soliton mass.

For the corroboration of the results obtained in continuous space–time, we have calculated the quantum corrections due to Gaussian fluctuations around a soliton within the discretisation technique with periodic and anti–periodic boundary conditions and rigid walls. We have proofed that the results coincide completely with those obtained in continuous space–time and are independent on the boundary conditions. Finally, it was shown that the finite contribution to the quantum mass of a soliton found in the literature arises due to a non–covariant procedure.

A. Thirring Model

A.1. Generating functional of Green functions

A.1.1. Evaluation of the Green function

In this section we will solve the Green function obeying the equation

$$i\gamma^\mu \left(\partial_\mu - iA_\mu(x)\right) S_A(x,y) = -\delta^{(2)}(x-y). \tag{A.1}$$

First we will prove that the solution is of the general form

$$S_A(x,y) = S_0(x,y) e^{i(\phi(x)-\phi(y))}, \tag{A.2}$$

where $S_0(x,y)$ is the free Green function (see Ref. [8]) obeying

$$i\gamma^\mu \partial_\mu S_0(x,y) = -\delta^{(2)}(x-y), \tag{A.3}$$

and $\phi(x)$ contains only parts proportional to the unit matrix and γ^5. Due to the absence of terms proportional to γ^μ in $\phi(x)$, derivatives of the Green function $S_A(x,y)$ can be performed trivially.

Acting from the left and from the right with γ^5 on Eq. (A.1) and Eq. (A.3) we find the symmetries

$$\begin{aligned}\gamma^5 S_A(x,y)\gamma^5 &= -S_A(x,y),\\ \gamma^5 S_0(x,y)\gamma^5 &= -S_0(x,y),\end{aligned} \tag{A.4}$$

and thus

$$\gamma^5 e^{i(\phi(x)-\phi(y))} \gamma^5 = e^{i(\phi(x)-\phi(y))}. \tag{A.5}$$

Due to the relations

$$\gamma^\mu \gamma^\nu = g^{\mu\nu} + \varepsilon^{\mu\nu}\gamma^5, \tag{A.6}$$
$$\gamma^\mu \gamma^5 = -\varepsilon^{\mu\nu}\gamma_\nu, \tag{A.7}$$

which are easy to verify, the exponent can be written in the form

$$i(\phi(x) - \phi(y)) = A + B\gamma^5 + C^\mu \gamma_\mu, \tag{A.8}$$

with A, B and C^μ being functions of x and y. Since

$$\gamma^0 = \sigma_1,$$
$$\gamma^1 = -i\sigma_2,$$
$$\gamma^5 = \sigma_3, \quad \text{(A.9)}$$

with Pauli matrices σ_i, the exponent can be rewritten as

$$\phi(x) - \phi(y) = \alpha + \beta \vec{n}\vec{\sigma},$$
$$e^{i(\phi(x)-\phi(y))} = e^{i\alpha}(\cos\beta + i\vec{n}\vec{\sigma}\sin\beta), \quad \text{(A.10)}$$

with $\vec{n}^2 = 1$ and α, β and $\vec{n}$ being again functions of x and y. Eq. (A.5) yields

$$\gamma^5 \vec{n}\vec{\sigma} \gamma^5 = \vec{n}\vec{\sigma}, \quad \text{(A.11)}$$

respectively

$$\sigma_3 \vec{n}\vec{\sigma} \sigma_3 = \vec{n}\vec{\sigma}, \quad \text{(A.12)}$$

due to Eq. (A.9). Using $\sigma_i \sigma_j = \delta_{ij} + i\varepsilon_{ijk}\sigma_k$ we have

$$\sigma_3 \sigma_i \sigma_3 = (\delta_{3i} + i\varepsilon_{3ij}\sigma_j)\sigma_3$$
$$= \delta_{3i}\sigma_3 + i\varepsilon_{3ij}(\delta_{j3} + i\varepsilon_{j3k}\sigma_k)$$
$$= \delta_{3i}\sigma_3 - (\delta_{33}\delta_{ik} - \delta_{3k}\delta_{i3})\sigma_k$$
$$= 2\delta_{3i}\sigma_3 - \sigma_i, \quad \text{(A.13)}$$

and

$$2n_3\sigma_3 - \vec{n}\vec{\sigma} = \vec{n}\vec{\sigma}, \quad \text{(A.14)}$$

yielding $n_1 = n_2 = 0$ and due to Eq. (A.9)

$$S_A(x,y) = S_0(x,y)e^{i(A+B\gamma^5)}. \quad \text{(A.15)}$$

Using Eq. (A.15) with Eq. (A.1) and Eq. (A.3) yields

$$i\gamma^\mu S_0(x,y)\partial_\mu e^{i(A+B\gamma^5)} + \gamma^\mu A_\mu(x) S_0(x,y) e^{i(A+B\gamma^5)}, \quad \text{(A.16)}$$

respectively

$$\gamma^\mu S_0(x,y)\partial_\mu(A+B\gamma^5) = \gamma^\mu A_\mu(x) S_0(x,y). \quad \text{(A.17)}$$

A.1. Generating functional of Green functions

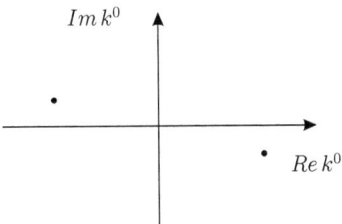

Figure A.1.: The poles in the complex k^0–plane of the causal $\Delta_F(x-y;\mu)$–function.

Due to Eq. (A.4) we have

$$\gamma^\mu \partial_\mu (A - B\gamma^5) S_0(x,y) = \gamma^\mu A_\mu(x) S_0(x,y) \,, \tag{A.18}$$

respectively

$$\gamma^\mu \partial_\mu (A - B\gamma^5) = \gamma^\mu A_\mu(x) \,. \tag{A.19}$$

Obviously, the solution is given by

$$A - B\gamma^5 = -i \int d^2z \left(S_0(x,z) - S_0(y,z) \right) \gamma^\mu A_\mu(z) \,. \tag{A.20}$$

Next we evaluate $S_0(x,y)$. Acting with $-i\gamma^\nu \partial_\nu$ on Eq. (A.3) yields

$$\Box S_0(x,y) = i\gamma^\mu \partial_\mu \delta^{(2)}(x-y) \,,$$

$$S_0(x,y) = i\gamma^\mu \partial_\mu \int \frac{d^2k}{(2\pi)^2} \frac{e^{-ik(x-y)}}{-k^2 - i0} \,. \tag{A.21}$$

Next we perform the integral. Introducing a small regularising mass μ we have (see Fig. A.1)

$$\int \frac{d^2k}{(2\pi)^2} \frac{e^{-ik(x-y)}}{-k^2 + \mu^2 - i0} = \int \frac{dk^1}{2\pi} \frac{e^{ik^1(x^1-y^1)}}{2\bar{k}} \int \frac{dk^0}{2\pi} e^{-ik^0(x^0-y^0)} \left(\frac{1}{k^0 + \bar{k}} - \frac{1}{k^0 - \bar{k}} \right)$$

$$= \int \frac{dk^1}{2\pi} \frac{e^{ik^1(x^1-y^1)}}{2\bar{k}} \left[i\Theta(x^0-y^0) e^{-i\bar{k}(x^0-y^0)} + i\Theta(y^0-x^0) e^{i\bar{k}(x^0-y^0)} \right]$$

$$= -\int \frac{dk^1}{2\pi i} \frac{e^{ik^1(x^1-y^1)}}{2\bar{k}} e^{-i\bar{k}|x^0-y^0|}$$

$$= -\int \frac{dk}{4\pi i} \frac{e^{ik\mu(x^1-y^1)}}{\sqrt{k^2+1-i0}} e^{-i\mu\sqrt{k^2+1-i0}|x^0-y^0|} \,, \tag{A.22}$$

with $\bar{k} = \sqrt{k_1^2 + \mu^2 - i0}$. Substitution of $k = \sinh t$ yields

$$-\int \frac{dk}{4\pi i} \frac{e^{ik\mu(x^1-y^1)}}{\sqrt{k^2+1-i0}} e^{-i\mu\sqrt{k^2+1-i0}|x^0-y^0|} =$$

$$\begin{aligned}
&= -\frac{1}{4\pi i}\int dt\, e^{i\sinh t\mu(x^1-y^1)-i\mu\cosh t|x^0-y^0|(1-i0)}\\
&= -\frac{1}{4\pi i}\int dt\, e^{-z\cosh(t+\tau)}\\
&= -\frac{1}{4\pi i}\int dt\, e^{-z\cosh t}\\
&= -\frac{1}{4\pi i}\left[\int_0^\infty dt\, e^{-z\cosh t} + \int_{-\infty}^0 dt\, e^{-z\cosh t}\right]\\
&= -\frac{1}{2\pi i}\int_0^\infty dt\, e^{-z\cosh t},
\end{aligned} \qquad (A.23)$$

where we have used

$$\cosh(t+\tau) = \sinh t \sinh\tau + \cosh t \cosh\tau. \qquad (A.24)$$

We find

$$\begin{aligned}
z\sinh\tau &= -i\mu(x^1-y^1),\\
z\cosh\tau &= i\mu|x^0-y^0|(1-i0),
\end{aligned} \qquad (A.25)$$

respectively

$$\begin{aligned}
z^2 &= [i\mu|x^0-y^0|(1-i0)]^2 - [-i\mu(x^1-y^1)]^2\\
&= -\mu^2(x-y)^2 + i0.
\end{aligned} \qquad (A.26)$$

Due to [83] (p.376, 9.6.23)

$$K_0(z) = \int_0^\infty e^{-z\cosh t}dt \qquad (|\arg z| < \pi/2), \qquad (A.27)$$

we arrive at

$$\int \frac{d^2k}{(2\pi)^2}\frac{e^{-ik(x-y)}}{-k^2+\mu^2-i0} = -\frac{1}{2\pi i}K_0\left(\sqrt{-\mu^2(x-y)^2+i0}\right). \qquad (A.28)$$

Since

$$K_0(z) = -\gamma + \ln(2) - \ln(z) + \mathcal{O}(z^2), \qquad (A.29)$$

(see "Mathematica" or [83] (p.375, 9.6.8)) where $\gamma = 0.577216$ is Euler's constant, we find (constant terms are dropped)

$$\begin{aligned}
\int \frac{d^2k}{(2\pi)^2}\frac{e^{-ik(x-y)}}{-k^2+\mu^2-i0} &= \frac{1}{4\pi i}\ln[-\mu^2(x-y)^2+i0]\\
&= \Delta_F(x-y;\mu).
\end{aligned} \qquad (A.30)$$

A.1. Generating functional of Green functions

Thus we obtain

$$S_0(x,y) = i\gamma^\mu \partial_\mu \Delta_F(x-y;\mu)$$
$$= \frac{1}{2\pi} \frac{\gamma^\mu(x-y)_\mu}{(x-y)^2 - i0}. \tag{A.31}$$

Using Eqs. (A.31), (A.20), (A.6) and (A.15) yields

$$S_A(x,y) = \frac{1}{2\pi} \frac{\gamma^\mu(x-y)_\mu}{(x-y)^2 - i0}$$
$$\times \exp\left\{-i(g^{\alpha\beta} - \varepsilon^{\alpha\beta}\gamma^5)\int d^2z \frac{\partial}{\partial z^\alpha}[\Delta_F(x-z;\mu) - \Delta_F(y-z;\mu)]A_\beta(z)\right\}, \tag{A.32}$$

where we have interchanged the derivative with respect to x respectively y with a derivative with respect to z.

A.1.2. Evaluation and regularisation of the vector current

The current we define as

$$j^\mu(x) = \lim_{y\to x} \bar\psi(y)\gamma^\mu \exp\left\{i\int_x^y dz^\nu \left[aA_\nu(z) + b\gamma^5 A_{5\nu}(z) + c\int d^2t \frac{\partial}{\partial t^\nu}\frac{\partial}{\partial t_\beta}\Delta_F(z-t;\mu)A_\beta(t)\right.\right.$$
$$\left.\left.+ d\gamma^5\int d^2t \frac{\partial}{\partial t^\nu}\frac{\partial}{\partial t_\beta}\Delta_F(z-t;\mu)A_{5\beta}(t)\right]\right\}\psi(x). \tag{A.33}$$

Under the gauge transformation $A_\nu \to A'_\nu = A_\nu + \partial_\nu \phi$ the third term in Eq. (A.33) behaves like the first one, whereas the fourth one is gauge invariant. In terms of the generating functional we find

$$j^\mu(x) = \lim_{y\to x} \frac{1}{Z_{\text{Th}}^{(0)}[A;J,\bar J]} \text{Tr}\left\{\left(i\frac{\delta}{\delta J(y)}\right)\gamma^\mu \exp\left\{i\int_x^y dz^\nu \left[aA_\nu(z) + b\gamma^5 A_{5\nu}(z)\right.\right.\right.$$
$$\left.\left.+ c\int d^2t \frac{\partial}{\partial t^\nu}\frac{\partial}{\partial t_\beta}\Delta_F(z-t;\mu)A_\beta(t) + d\gamma^5\int d^2t \frac{\partial}{\partial t^\nu}\frac{\partial}{\partial t_\beta}\Delta_F(z-t;\mu)A_{5\beta}(t)\right]\right\}$$
$$\left.\times\left(\frac{1}{i}\frac{\delta}{\delta \bar J(x)}\right)\right\}Z_{\text{Th}}^{(0)}[A;J,\bar J]\bigg|_{J=\bar J=0}. \tag{A.34}$$

With the relation

$$\langle j^\mu(x)\rangle_A = \frac{\langle 0|j^\mu(x)|0\rangle_{A;g=J=\bar J=0}}{Z_{\text{Th}}^{(0)}[A;0,0]}, \tag{A.35}$$

we have

$$\langle j^\mu(x)\rangle_A = \lim_{y\to x}\text{Tr}\left\{iS_A(x,y)\gamma^\mu \exp\left\{i\int_x^y dz^\nu\left[aA_\nu(z) + b\gamma^5 A_{5\nu}(z)\right.\right.\right.$$
$$\left.\left.\left.+ c\int d^2t \frac{\partial}{\partial t^\nu}\frac{\partial}{\partial t_\beta}\Delta_F(z-t;\mu)A_\beta(t) + d\gamma^5\int d^2t \frac{\partial}{\partial t^\nu}\frac{\partial}{\partial t_\beta}\Delta_F(z-t;\mu)A_{5\beta}(t)\right]\right\}\right\}. \tag{A.36}$$

The average of the vector current is given by

$$\langle j^\mu(x)\rangle_A = \lim_{y\to x} \frac{i}{2\pi} \frac{(x-y)_\rho}{(x-y)^2 - i0} \mathrm{Tr}\bigg\{\gamma^\rho \exp\Big[-i(g^{\alpha\beta} - \varepsilon^{\alpha\beta}\gamma^5)$$
$$\times \int d^2z \frac{\partial}{\partial z^\alpha}[\Delta_F(x-z;\mu) - \Delta_F(y-z;\mu)]A_\beta(z)\Big]\gamma^\mu$$
$$\times \exp\bigg\{i\int_x^y dz^\nu \Big[aA_\nu(z) + b\gamma^5 A_{5\nu}(z) + c\int d^2t \frac{\partial}{\partial t^\nu}\frac{\partial}{\partial t_\beta}\Delta_F(z-t;\mu)A_\beta(t)$$
$$+ d\gamma^5 \int d^2t \frac{\partial}{\partial t^\nu}\frac{\partial}{\partial t_\beta}\Delta_F(z-t;\mu)A_{5\beta}(t)\Big]\bigg\}\bigg\}. \qquad (A.37)$$

Performing the limit $y^0 = x^0$, $\epsilon \to 0$ with $y^1 = x^1 \pm \epsilon$ we find

$$\langle j^\mu(x)\rangle_A = \lim_{\epsilon\to 0} \frac{i}{2\pi}\frac{1}{\mp\epsilon}\mathrm{Tr}\bigg\{\gamma^1\Big[1 \mp i\epsilon(g^{\alpha\beta} - \varepsilon^{\alpha\beta}\gamma^5)\frac{\partial}{\partial x^1}\frac{\partial}{\partial x^\alpha}\int d^2z \,\Delta_F(x-z;\mu)A_\beta(z)\Big]$$
$$\times \gamma^\mu\Big[1 \pm i\epsilon\Big(aA_1(x) + b\gamma^5 A_{51}(x) + c\int d^2t \frac{\partial}{\partial t^1}\frac{\partial}{\partial t^\beta}\Delta_F(x-t;\mu)A^\beta(t)$$
$$+ d\gamma^5 \int d^2t \frac{\partial}{\partial t^1}\frac{\partial}{\partial t^\beta}\Delta_F(x-t;\mu)A_5^\beta(t)\Big)\Big]\bigg\}$$
$$= \mp\lim_{\epsilon\to 0}\frac{ig^{1\mu}}{\pi\epsilon} \mp \lim_{\epsilon\to 0}\frac{i}{2\pi\epsilon}\Big[\mp i\epsilon(2g^{1\mu}g^{\alpha\beta} + 2\varepsilon^{1\mu}\varepsilon^{\alpha\beta})\frac{\partial}{\partial x^1}\frac{\partial}{\partial x^\alpha}\int d^2z \,\Delta_F(x-z;\mu)A_\beta(z)$$
$$\pm i\epsilon\Big(2ag^{1\mu}A_1(x) + 2b\varepsilon^{1\mu}A_{51}(x) + 2cg^{1\mu}\int d^2t \frac{\partial}{\partial t^1}\frac{\partial}{\partial t^\beta}\Delta_F(x-t;\mu)A^\beta(t)$$
$$+ 2d\varepsilon^{1\mu}\int d^2t \frac{\partial}{\partial t^1}\frac{\partial}{\partial t^\beta}\Delta_F(x-t;\mu)A_5^\beta(t)\Big)\Big], \qquad (A.38)$$

where we have used

$$\mathrm{Tr}(\gamma^1\gamma^\mu) = 2g^{1\mu}, \qquad \mathrm{Tr}(\gamma^1\gamma^\mu\gamma^5) = \mathrm{Tr}((g^{1\mu} + \varepsilon^{1\mu}\gamma^5)\gamma^5) = 2\varepsilon^{1\mu}. \qquad (A.39)$$

Performing the symmetric limit we arrive at

$$\langle j^\mu(x)\rangle_A = \frac{1}{\pi}\bigg[-(g^{1\mu}g^{\alpha\beta} + \varepsilon^{1\mu}\varepsilon^{\alpha\beta})\frac{\partial}{\partial x^1}\frac{\partial}{\partial x^\alpha}\int d^2z \,\Delta_F(x-z;\mu)A_\beta(z)$$
$$+ \Big(ag^{1\mu}A_1(x) + b\varepsilon^{1\mu}A_{51}(x) + cg^{1\mu}\int d^2t \frac{\partial}{\partial t^1}\frac{\partial}{\partial t^\beta}\Delta_F(x-t;\mu)A^\beta(t)$$
$$+ d\varepsilon^{1\mu}\int d^2t \frac{\partial}{\partial t^1}\frac{\partial}{\partial t^\beta}\Delta_F(x-t;\mu)A_5^\beta(t)\Big)\bigg]. \qquad (A.40)$$

The components of the current read

$$\langle j^0(x)\rangle_A = \frac{1}{\pi}\varepsilon^{\alpha\beta}\frac{\partial}{\partial x^1}\frac{\partial}{\partial x^\alpha}\int d^2z \,\Delta_F(x-z;\mu)A_\beta(z) - \frac{b}{\pi}A_{51}(x)$$
$$- \frac{d}{\pi}\int d^2t \frac{\partial}{\partial t^1}\frac{\partial}{\partial t^\beta}\Delta_F(x-t;\mu)A_5^\beta(t), \qquad (A.41)$$

A.1. Generating functional of Green functions

and

$$\langle j^1(x)\rangle_A = \frac{1}{\pi}\frac{\partial}{\partial x^1}\frac{\partial}{\partial x^\alpha}\int d^2z\,\Delta_F(x-z;\mu)A^\alpha(z) - \frac{a}{\pi}A_1(x)$$
$$- \frac{c}{\pi}\int d^2t\frac{\partial}{\partial t^1}\frac{\partial}{\partial t^\beta}\Delta_F(x-t;\mu)A^\beta(t)\,. \tag{A.42}$$

Using $A_{5\mu} = -\varepsilon_{\mu\nu}A^\nu$ and $\Box\Delta_F(x-y;\mu) = \delta^{(2)}(x-y)$ the zero component can be rewritten as follows

$$\langle j^0(x)\rangle_A = \frac{1}{\pi}\frac{\partial}{\partial x^1}\frac{\partial}{\partial x^0}\int d^2z\,\Delta_F(x-z;\mu)A_1(z) - \frac{1}{\pi}\frac{\partial}{\partial x^1}\frac{\partial}{\partial x^1}\int d^2z\,\Delta_F(x-z;\mu)A_0(z)$$
$$+ \frac{b}{\pi}A^0(x) + \frac{d}{\pi}\int d^2t\frac{\partial}{\partial t^1}\frac{\partial}{\partial t^0}\Delta_F(x-t;\mu)A_1(t) - \frac{d}{\pi}\int d^2t\frac{\partial}{\partial t^1}\frac{\partial}{\partial t^1}\Delta_F(x-t;\mu)A_0(t)$$
$$= -\frac{1}{\pi}\frac{\partial}{\partial x^0}\frac{\partial}{\partial x^\mu}\int d^2z\,\Delta_F(x-z;\mu)A^\mu(z) + \frac{1+b}{\pi}A^0(x)$$
$$- \frac{d}{\pi}\int d^2t\frac{\partial}{\partial t^0}\frac{\partial}{\partial t^\mu}\Delta_F(x-t;\mu)A^\mu(t) + \frac{d}{\pi}A^0(x)$$
$$= -\frac{d+1}{\pi}\frac{\partial}{\partial x_0}\frac{\partial}{\partial x^\mu}\int d^2z\,\Delta_F(x-z;\mu)A^\mu(z) + \frac{b+d+1}{\pi}A^0(x)\,. \tag{A.43}$$

Comparison with the one component

$$\langle j^1(x)\rangle_A = \frac{c-1}{\pi}\frac{\partial}{\partial x_1}\frac{\partial}{\partial x^\mu}\int d^2z\,\Delta_F(x-z;\mu)A^\mu(z) + \frac{a}{\pi}A^1(x)\,, \tag{A.44}$$

yields that covariance is obtained only for $c = -d$ and $b+d+1 = a$. Hence the two-current can be written as

$$\langle j^\mu(x)\rangle_A = -\frac{\chi}{\pi}\frac{\partial}{\partial x_\mu}\frac{\partial}{\partial x^\nu}\int d^2z\,\Delta_F(x-z;\mu)A^\nu(z) + \frac{\zeta}{\pi}A^\mu(x)\,, \tag{A.45}$$

with two arbitrary parameters χ and ζ. Introducing

$$D^{(0)}_{\mu\nu}(x-y) = \frac{\zeta}{\pi}g_{\mu\nu}\delta^{(2)}(x-y) - \frac{\chi}{\pi}\frac{\partial}{\partial x^\mu}\frac{\partial}{\partial x^\nu}\Delta_F(x-y;\mu)\,, \tag{A.46}$$

we can write the average of the current as

$$\langle j_\mu(x)\rangle_A = \int d^2z\,D^{(0)}_{\mu\nu}(x-z)A^\nu(z)\,. \tag{A.47}$$

A.1.3. Evaluation of the derivative of the vector current

The derivative of the current is given by

$$\frac{\delta j^\mu(x)}{\delta A_\nu(y)} = \lim_{z\to x}\bar\psi(z)\gamma^\mu i\int_x^z d\sigma^\lambda\left[ag^\nu_\lambda\delta^{(2)}(\sigma-y) - b\gamma^5\varepsilon_\lambda^{\,\nu}\delta^{(2)}(\sigma-y) + c\frac{\partial}{\partial\sigma^\lambda}\frac{\partial}{\partial\sigma_\nu}\Delta_F(\sigma-y;\mu)\right.$$
$$\left. - d\gamma^5\varepsilon_\beta^{\,\nu}\frac{\partial}{\partial\sigma^\lambda}\frac{\partial}{\partial\sigma_\beta}\Delta_F(\sigma-y;\mu)\right]$$

$$\times \exp\left\{i\int_x^z d\sigma^\lambda \left[aA_\lambda(\sigma) + b\gamma^5 A_{5\lambda}(\sigma) + c\int d^2t \frac{\partial}{\partial t^\lambda}\frac{\partial}{\partial t_\beta}\Delta_F(\sigma-t;\mu)A_\beta(t)\right.\right.$$
$$\left.\left.+ d\gamma^5 \int d^2t \frac{\partial}{\partial t^\lambda}\frac{\partial}{\partial t_\beta}\Delta_F(\sigma-t;\mu)A_{5\beta}(t)\right]\right\}\psi(x)\,, \qquad (A.48)$$

and its average reads

$$\langle \frac{\delta j^\mu(x)}{\delta A_\nu(y)}\rangle_A = \lim_{z\to x}\frac{i}{2\pi}\frac{(x-z)_\rho}{(x-z)^2-i0}\mathrm{Tr}\left\{\gamma^\rho \exp\left[-i(g^{\alpha\beta}-\varepsilon^{\alpha\beta}\gamma^5)\right.\right.$$
$$\times \int d^2\sigma \frac{\partial}{\partial \sigma^\alpha}[\Delta_F(x-\sigma;\mu)-\Delta_F(z-\sigma;\mu)]A_\beta(\sigma)\bigg]\gamma^\mu$$
$$\times i\int_x^z d\sigma^\lambda\left[ag_\lambda^\nu\delta^{(2)}(\sigma-y) - b\gamma^5 \varepsilon_\lambda^{\ \nu}\delta^{(2)}(\sigma-y) + c\frac{\partial}{\partial \sigma^\lambda}\frac{\partial}{\partial \sigma_\nu}\Delta_F(\sigma-y;\mu)\right.$$
$$\left. - d\gamma^5 \varepsilon_\beta^{\ \nu}\frac{\partial}{\partial \sigma^\lambda}\frac{\partial}{\partial \sigma_\beta}\Delta_F(\sigma-y;\mu)\right]$$
$$\times \exp\left\{i\int_x^z d\sigma^\lambda\left[aA_\lambda(\sigma) + b\gamma^5 A_{5\lambda}(\sigma) + c\int d^2t \frac{\partial}{\partial t^\lambda}\frac{\partial}{\partial t_\beta}\Delta_F(\sigma-t;\mu)A_\beta(t)\right.\right.$$
$$\left.\left.\left.+ d\gamma^5 \int d^2t \frac{\partial}{\partial t^\lambda}\frac{\partial}{\partial t_\beta}\Delta_F(\sigma-t;\mu)A_{5\beta}(t)\right]\right\}\right\}$$
$$= \lim_{\epsilon\to 0}\frac{i}{2\pi}\frac{1}{\mp\epsilon}\mathrm{Tr}\left\{\pm i\epsilon\gamma^1\gamma^\mu\left(ag_1^\nu\delta^{(2)}(x-y) - b\gamma^5\varepsilon_1^{\ \nu}\delta^{(2)}(x-y)\right.\right.$$
$$\left.\left.+ c\frac{\partial}{\partial x^1}\frac{\partial}{\partial x_\nu}\Delta_F(x-y;\mu) - d\gamma^5\varepsilon_\beta^{\ \nu}\frac{\partial}{\partial x^1}\frac{\partial}{\partial x_\beta}\Delta_F(x-y;\mu)\right)\right\}$$
$$= \frac{a}{\pi}g^{1\mu}g_1^\nu \delta^{(2)}(x-y) - \frac{b}{\pi}\varepsilon^{1\mu}\varepsilon_1^{\ \nu}\delta^{(2)}(x-y) + \frac{c}{\pi}g^{1\mu}\frac{\partial}{\partial x^1}\frac{\partial}{\partial x_\nu}\Delta_F(x-y;\mu)$$
$$- \frac{d}{\pi}\varepsilon^{1\mu}\varepsilon_\beta^{\ \nu}\frac{\partial}{\partial x^1}\frac{\partial}{\partial x_\beta}\Delta_F(x-y;\mu)\,. \qquad (A.49)$$

Using the relations

$$a = \zeta\,,$$
$$b = \zeta - \chi\,,$$
$$c = 1 - \chi\,,$$
$$d = \chi - 1\,, \qquad (A.50)$$

we find for the component $\mu = \nu = 0$

$$\langle \frac{\delta j^0(x)}{\delta A_0(y)}\rangle_A = \frac{\zeta-\chi}{\pi}\delta^{(2)}(x-y) + \frac{\chi-1}{\pi}\frac{\partial}{\partial x^1}\frac{\partial}{\partial x_1}\Delta_F(x-y;\mu)$$
$$= \frac{\zeta-1}{\pi}\delta^{(2)}(x-y) - \frac{\chi-1}{\pi}\frac{\partial}{\partial x^0}\frac{\partial}{\partial x_0}\Delta_F(x-y;\mu)\,, \qquad (A.51)$$

where we have used $\Box\Delta_F(x-y;\mu) = \delta^{(2)}(x-y)$. The component $\mu=0, \nu=1$ reads

$$\langle \frac{\delta j^0(x)}{\delta A_1(y)}\rangle_A = \frac{\chi-1}{\pi}\frac{\partial}{\partial x^1}\frac{\partial}{\partial x_0}\Delta_F(x-y;\mu)\,. \qquad (A.52)$$

A.1. Generating functional of Green functions

For the $\mu = 1, \nu = 0$ component we find

$$\langle \frac{\delta j^1(x)}{\delta A_0(y)} \rangle_A = \frac{\chi - 1}{\pi} \frac{\partial}{\partial x^1} \frac{\partial}{\partial x_0} \Delta_F(x-y;\mu), \tag{A.53}$$

while for $\mu = \nu = 1$ we have

$$\langle \frac{\delta j^1(x)}{\delta A_1(y)} \rangle_A = -\frac{\zeta}{\pi} \delta^{(2)}(x-y) + \frac{\chi - 1}{\pi} \frac{\partial}{\partial x^1} \frac{\partial}{\partial x_1} \Delta_F(x-y;\mu). \tag{A.54}$$

Hence we find

$$\langle \frac{\delta j_\mu(x)}{\delta A^\nu(y)} \rangle_A = \left(-\frac{1}{\pi} g_{\mu 0} g_{\nu 0} + \frac{\zeta}{\pi} g_{\mu\nu} \right) \delta^{(2)}(x-y) - \frac{\chi - 1}{\pi} \frac{\partial}{\partial x^\mu} \frac{\partial}{\partial x^\nu} \Delta_F(x-y;\mu)$$

$$= \frac{1}{\pi} \left(-g_{\mu 0} g_{\nu 0} \delta^{(2)}(x-y) + \frac{\partial}{\partial x^\mu} \frac{\partial}{\partial x^\nu} \Delta_F(x-y;\mu) \right) + D^{(0)}_{\mu\nu}(x-y). \tag{A.55}$$

From Eqs. (A.48) and (A.49) one can read off that the average of all functional derivatives of second order or higher of the vector current j^μ with respect to the external vector source A^ν vanish, i.e.

$$\langle \frac{\delta^n j^\mu(x)}{\delta A_{\nu_1}(z_1) \dots \delta A_{\nu_n}(z_n)} \rangle_A = \lim_{\epsilon \to 0} \mathcal{O}(\epsilon^{n-1})$$

$$= 0 \qquad \text{for} \quad n \geq 2, \tag{A.56}$$

where $n \in \mathbb{N}$.

A.1.4. Performing the path integral

Using that $\int \mathcal{D}^2 u \, e^{uLu} = 1$ with appropriately chosen integral measure and the invariance of the path integral under a shift $u \to u + \Delta$ we find in "matrix–notation"

$$\int \mathcal{D}^2 u \exp\left\{ uLu + 2\Delta Lu \right\} = \exp\left\{ -\Delta L \Delta \right\}. \tag{A.57}$$

For $L = -\frac{i}{2}$ and $\Delta = -\sqrt{g} \frac{\delta}{\delta A_\mu(x)}$ we have

$$\exp\left\{ \frac{i}{2} g \int d^2 x \frac{\delta}{\delta A_\mu(x)} \frac{\delta}{\delta A^\mu(x)} \right\} = \int \mathcal{D}^2 u \exp\left\{ -\frac{i}{2} \int d^2 z \, u_\mu(z) u^\mu(z) + i \sqrt{g} \int d^2 z \, u_\mu(z) \frac{\delta}{\delta A_\mu(z)} \right\}. \tag{A.58}$$

A.1.5. Evaluation of $Z^{(g)}_{\text{Th}}[A; 0, 0]$

In this subsection we evaluate $Z^{(g)}_{\text{Th}}[A; 0, 0]$, this means interaction is turned on. We find

$$Z^{(g)}_{\text{Th}}[A; 0, 0] = \exp\left\{ \frac{i}{2} g \int d^2 x \frac{\delta}{\delta A_\mu(x)} \frac{\delta}{\delta A^\mu(x)} \right\} \exp\left\{ \frac{i}{2} \int d^2 x d^2 y \, A^\mu(x) D^{(0)}_{\mu\nu}(x-y) A^\nu(y) \right\}$$

$$\begin{aligned}
&= \exp\left\{\frac{i}{2}g\int d^2x \frac{\delta}{\delta J_\mu(x)}\frac{\delta}{\delta J^\mu(x)}\right\} \\
&\quad \times \exp\left\{\frac{i}{2}\int d^2x d^2y \left(J^\mu(x)+A^\mu(x)\right)D^{(0)}_{\mu\nu}(x-y)\left(J^\nu(y)+A^\nu(y)\right)\right\}\bigg|_{J=0} \\
&= \int \mathcal{D}^2 u \exp\left\{-\frac{i}{2}\int d^2x d^2y\, u^\mu(x)\left(g_{\mu\nu}\delta^{(2)}(x-y)+gD^{(0)}_{\mu\nu}(x-y)\right)u^\nu(y)\right. \\
&\quad \left. -\sqrt{g}\int d^2x d^2y\, A^\mu(x)D^{(0)}_{\mu\nu}(x-y)u^\nu(y)+\frac{i}{2}\int d^2x d^2y\, A^\mu(x)D^{(0)}_{\mu\nu}(x-y)A^\nu(y)\right\}.
\end{aligned}$$
(A.59)

Using Eq. (A.57) with

$$L_{\mu\nu} = -\frac{i}{2}(1+gD)_{\mu\nu},$$
$$2\Delta^\mu L_{\mu\nu} = -\sqrt{g}\, A^\mu D_{\mu\nu},$$
$$\Delta^\mu = -i\sqrt{g}\, A^\nu D_{\nu\sigma}((1+gD)^{-1})^{\sigma\mu},$$
(A.60)

and

$$-\Delta L \Delta = -\frac{i}{2} g\, AD\frac{1}{1+gD}DA,$$
(A.61)

we have

$$Z^{(g)}_{\text{Th}}[A;0,0] = \exp\left\{\frac{i}{2}A(D - gD\frac{1}{1+gD}D)A\right\}.$$
(A.62)

Since

$$\begin{aligned}
D - gD\frac{1}{1+gD}D &= D\frac{1}{1+gD}(1+gD) - gD\frac{1}{1+gD}D \\
&= D\frac{1}{1+gD},
\end{aligned}$$
(A.63)

we arrive at

$$Z^{(g)}_{\text{Th}}[A;0,0] = \exp\left\{\frac{i}{2}AD\frac{1}{1+gD}A\right\}.$$
(A.64)

Next we evaluate $D\dfrac{1}{1+gD}$.

$$\begin{aligned}
D\frac{1}{1+gD} &= \int d^2z \left(\frac{\zeta}{\pi}g^{\mu\alpha}\delta^{(2)}(x-z) - \frac{\chi}{\pi}\frac{\partial}{\partial x_\mu}\frac{\partial}{\partial x_\alpha}\Delta_F(x-z;\mu)\right) \\
&\quad \times \left(\frac{g^\nu_\alpha}{1+\zeta\frac{g}{\pi}}\delta^{(2)}(z-y) + \frac{g}{\pi}\frac{\chi}{\left(1+\zeta\frac{g}{\pi}\right)\left(1+(\zeta-\chi)\frac{g}{\pi}\right)}\frac{\partial}{\partial z^\alpha}\frac{\partial}{\partial z_\nu}\Delta_F(z-y;\mu)\right) \\
&= \frac{\zeta}{\pi}\frac{g^{\mu\nu}}{1+\zeta\frac{g}{\pi}}\delta^{(2)}(x-y) - \frac{\chi}{\pi}\frac{1}{1+\zeta\frac{g}{\pi}}\frac{\partial}{\partial x_\mu}\frac{\partial}{\partial x_\nu}\Delta_F(x-y;\mu).
\end{aligned}$$

$$+ \frac{\zeta\chi}{\pi^2} \frac{g}{\left(1+\zeta\frac{g}{\pi}\right)\left(1+(\zeta-\chi)\frac{g}{\pi}\right)} \frac{\partial}{\partial x_\mu} \frac{\partial}{\partial x_\nu} \Delta_F(x-y;\mu)$$

$$- \frac{\chi^2}{\pi^2} \frac{g}{\left(1+\zeta\frac{g}{\pi}\right)\left(1+(\zeta-\chi)\frac{g}{\pi}\right)} \int d^2z \, \frac{\partial}{\partial x_\mu} \frac{\partial}{\partial x_\alpha} \Delta_F(x-z;\mu) \frac{\partial}{\partial z^\alpha} \frac{\partial}{\partial z_\nu} \Delta_F(z-y;\mu)$$

$$= \frac{\zeta}{\pi} \frac{g^{\mu\nu}}{1+\zeta\frac{g}{\pi}} \delta^{(2)}(x-y) + \left(-\frac{\chi}{\pi} \frac{1}{1+\zeta\frac{g}{\pi}} + \frac{\zeta\chi}{\pi^2} \frac{g}{\left(1+\zeta\frac{g}{\pi}\right)\left(1+(\zeta-\chi)\frac{g}{\pi}\right)} \right.$$

$$\left. - \frac{\chi^2}{\pi^2} \frac{g}{\left(1+\zeta\frac{g}{\pi}\right)\left(1+(\zeta-\chi)\frac{g}{\pi}\right)} \right) \frac{\partial}{\partial x_\mu} \frac{\partial}{\partial x_\nu} \Delta_F(x-y;\mu)$$

$$= \frac{\zeta}{\pi} \frac{g^{\mu\nu}}{1+\zeta\frac{g}{\pi}} \delta^{(2)}(x-y) - \frac{\chi}{\pi} \frac{1}{\left(1+\zeta\frac{g}{\pi}\right)\left(1+(\zeta-\chi)\frac{g}{\pi}\right)} \frac{\partial}{\partial x_\mu} \frac{\partial}{\partial x_\nu} \Delta_F(x-y;\mu)$$

$$= \frac{1}{\pi} \left[\frac{\zeta-\chi}{1+(\zeta-\chi)\frac{g}{\pi}} g^{\mu\alpha}g^{\nu\beta} - \frac{\zeta}{1+\zeta\frac{g}{\pi}} \varepsilon^{\mu\alpha}\varepsilon^{\nu\beta} \right] \frac{\partial}{\partial x^\alpha} \frac{\partial}{\partial x^\beta} \Delta_F(x-y;\mu), \quad \text{(A.65)}$$

where we have used $g^{\mu\nu}g^{\alpha\beta} = g^{\mu\beta}g^{\alpha\nu} - \varepsilon^{\mu\alpha}\varepsilon^{\nu\beta}$ and
$\Box \Delta_F(x-y;\mu) = \delta^{(2)}(x-y)$ in the last line. So we arrive at

$$Z^{(g)}_{\text{Th}}[A;0,0] = \exp\left\{ \frac{i}{2} \int d^2z_1 d^2z_2 \, A^\mu(z_1) D^{(g)}_{\mu\nu}(z_1-z_2) A^\nu(z_2) \right\}, \quad \text{(A.66)}$$

where we have introduced

$$D^{(g)}_{\mu\nu} = \frac{\zeta}{\pi} \frac{g_{\mu\nu}}{1+\zeta\frac{g}{\pi}} \delta^{(2)}(x-y) - \frac{\chi}{\pi} \frac{1}{\left(1+\zeta\frac{g}{\pi}\right)\left(1+(\zeta-\chi)\frac{g}{\pi}\right)} \frac{\partial}{\partial x^\mu} \frac{\partial}{\partial x^\nu} \Delta_F(x-y;\mu). \quad \text{(A.67)}$$

A.2. Two–point causal Green function $G(x,y)$

A.2.1. Evaluation of the Green function

The Green function reads

$$G(x,y) = \frac{1}{i} \frac{\delta}{\delta\bar{J}(x)} \frac{\delta}{\delta J(y)} Z^{(g)}_{\text{Th}}[A;J,\bar{J}] \bigg|_{A=J=\bar{J}=0}$$

$$= \exp\left\{ \frac{i}{2} g \int d^2z \frac{\delta}{\delta A_\mu(z)} \frac{\delta}{\delta A^\mu(z)} \right\}$$

$$\times \exp\left\{ \frac{i}{2} \int d^2z_1 d^2z_2 \, A^\mu(z_1) D^{(0)}_{\mu\nu}(z_1-z_2) A^\nu(z_2) \right\} S_A(x,y) \bigg|_{A=0}. \quad \text{(A.68)}$$

Using Eq. (A.58) the calculation of the Green function reduces to the calculation of the path integral

$$G(x,y) = \frac{1}{2\pi} \frac{\gamma^\mu(x-y)_\mu}{(x-y)^2 - i0} \int \mathcal{D}^2 u \exp\left\{ -\frac{i}{2} \int d^2z \, u_\mu(z) u^\mu(z) \right.$$

$$-\frac{i}{2}g\iint d^2z_1 d^2z_2\, u^\mu(z_1)\, D^{(0)}_{\mu\nu}(z_1 - z_2)\, u^\nu(z_2) + \sqrt{g}\,(g^{\alpha\beta} - \varepsilon^{\alpha\beta}\gamma^5)$$
$$\times \int d^2z\, \frac{\partial}{\partial z^\alpha}[\Delta_F(x - z;\mu) - \Delta_F(y - z;\mu)]\, u_\beta(z)\Big\},\qquad\text{(A.69)}$$

which we write as

$$G(x,y) = \frac{1}{2\pi}\frac{\gamma^\mu(x-y)_\mu}{(x-y)^2 - i0}$$
$$\times \int \mathcal{D}^2 u\, \exp\Big\{-\frac{i}{2}u(1+gD)u + \sqrt{g}\,\partial(\Delta_x - \Delta_y)\, u - \sqrt{g}\,\gamma^5\,\partial(\Delta_x - \Delta_y)\,\varepsilon\, u\Big\}.$$
(A.70)

Using Eq. (A.57) with

$$L_{\mu\nu} = -\frac{i}{2}(1+gD)_{\mu\nu}\,,$$
$$2\Delta^\mu L_{\mu\nu} = \sqrt{g}\,\partial(\Delta_x - \Delta_y)^\mu(1-\gamma^5\varepsilon)_{\mu\nu}\,,$$
$$\Delta^\mu = i\sqrt{g}\,\partial(\Delta_x - \Delta_y)^\nu(1-\gamma^5\varepsilon)_{\nu\rho}((1+gD)^{-1})^{\rho\mu}\,,\qquad\text{(A.71)}$$

and

$$-\Delta L \Delta = -\frac{i}{2}g\partial(\Delta_x - \Delta_y)(1-\gamma^5\varepsilon)\frac{1}{1+gD}(1+\gamma^5\varepsilon)\partial(\Delta_x - \Delta_y)$$
$$= -\frac{i}{2}g\partial(\Delta_x - \Delta_y)\frac{1}{1+gD}\partial(\Delta_x - \Delta_y) + \frac{i}{2}g\partial(\Delta_x - \Delta_y)\varepsilon\frac{1}{1+gD}\varepsilon\partial(\Delta_x - \Delta_y)$$
$$+ \frac{i}{2}g\gamma^5\partial(\Delta_x - \Delta_y)\varepsilon\frac{1}{1+gD}\partial(\Delta_x - \Delta_y) - \frac{i}{2}g\gamma^5\partial(\Delta_x - \Delta_y)\frac{1}{1+gD}\varepsilon\partial(\Delta_x - \Delta_y)\,,$$
(A.72)

we find

$$G(x,y) = \frac{1}{2\pi}\frac{\gamma^\mu(x-y)_\mu}{(x-y)^2 - i0}\exp\Big\{-\frac{i}{2}g\,\partial(\Delta_x - \Delta_y)\frac{1}{1+gD}\partial(\Delta_x - \Delta_y)$$
$$+ \frac{i}{2}g\,\partial(\Delta_x - \Delta_y)\,\varepsilon\,\frac{1}{1+gD}\,\varepsilon\,\partial(\Delta_x - \Delta_y) + \frac{i}{2}g\gamma^5\partial(\Delta_x - \Delta_y)\varepsilon\frac{1}{1+gD}\partial(\Delta_x - \Delta_y)$$
$$- \frac{i}{2}g\gamma^5\partial(\Delta_x - \Delta_y)\frac{1}{1+gD}\varepsilon\partial(\Delta_x - \Delta_y)\Big\}.\qquad\text{(A.73)}$$

For the subsequent calculations we have to evaluate the matrix element $\frac{1}{1+gD}$. The following relation holds

$$\int d^2z\,(1+gD^{(0)})^{\mu\alpha}(x,z)((1+gD^{(0)})^{-1})_{\alpha\nu}(z,y) = g^\mu_\nu\, \delta^{(2)}(x-y)\,,\qquad\text{(A.74)}$$

where $1+gD$ reads

$$(1+gD^{(0)})^{\mu\alpha}(x,z) = \Big(1+\zeta\frac{g}{\pi}\Big) g^{\mu\alpha}\,\delta^{(2)}(x-z) - \chi\frac{g}{\pi}\frac{\partial}{\partial x_\mu}\frac{\partial}{\partial x_\alpha}\Delta_F(x-z;\mu)\,.\qquad\text{(A.75)}$$

A.2. Two–point causal Green function $G(x,y)$

The ansatz

$$((1+gD^{(0)})^{-1})_{\alpha\nu}(z,y) = Ag_{\alpha\nu}\delta^{(2)}(z-y) + B\frac{\partial}{\partial z^\alpha}\frac{\partial}{\partial z^\nu}\Delta_F(z-y;\mu),\qquad (A.76)$$

yields

$$\int d^2z\,(1+gD^{(0)})^{\mu\alpha}(x,z)((1+gD^{(0)})^{-1})_{\alpha\nu}(z,y) =$$

$$= \int d^2z\left[\left(1+\zeta\frac{g}{\pi}\right)g^{\mu\alpha}\delta^{(2)}(x-z) - \chi\frac{g}{\pi}\frac{\partial}{\partial x_\mu}\frac{\partial}{\partial x_\alpha}\Delta_F(x-z;\mu)\right]$$

$$\times\left[Ag_{\alpha\nu}\delta^{(2)}(z-y) + B\frac{\partial}{\partial z^\alpha}\frac{\partial}{\partial z^\nu}\Delta_F(z-y;\mu)\right]$$

$$= \left(1+\zeta\frac{g}{\pi}\right)Ag^\mu_\nu\delta^{(2)}(x-y) - \chi\frac{g}{\pi}\frac{\partial}{\partial x_\mu}\frac{\partial}{\partial x^\nu}\Delta_F(x-y;\mu)A$$

$$+\left(1+\zeta\frac{g}{\pi}\right)B\frac{\partial}{\partial x_\mu}\frac{\partial}{\partial x^\nu}\Delta_F(x-y;\mu)$$

$$-\int d^2z\,\chi\frac{g}{\pi}\frac{\partial}{\partial x_\mu}\frac{\partial}{\partial x_\alpha}\Delta_F(x-z;\mu)B\frac{\partial}{\partial z^\alpha}\frac{\partial}{\partial z^\nu}\Delta_F(z-y;\mu)$$

$$= \left(1+\zeta\frac{g}{\pi}\right)Ag^\mu_\nu\delta^{(2)}(x-y) + \left(B+\zeta\frac{g}{\pi}B - \chi\frac{g}{\pi}A - \chi\frac{g}{\pi}B\right)\frac{\partial}{\partial x_\mu}\frac{\partial}{\partial x^\nu}\Delta_F(x-y;\mu),$$

$$(A.77)$$

where we have used $\Box\Delta_F(x-y;\mu) = \delta^{(2)}(x-y)$. Setting

$$\left(1+\zeta\frac{g}{\pi}\right)A = 1,$$

$$\left(B + \zeta\frac{g}{\pi}B - \chi\frac{g}{\pi}A - \chi\frac{g}{\pi}B\right) = 0,\qquad (A.78)$$

we arrive at

$$((1+gD^{(0)})^{-1})_{\alpha\nu}(z,y) =$$

$$= \frac{g_{\alpha\nu}}{1+\zeta\frac{g}{\pi}}\delta^{(2)}(z-y) + \frac{g}{\pi}\frac{\chi}{\left(1+\zeta\frac{g}{\pi}\right)\left(1+(\zeta-\chi)\frac{g}{\pi}\right)}\frac{\partial}{\partial z^\alpha}\frac{\partial}{\partial z^\nu}\Delta_F(z-y;\mu).\qquad (A.79)$$

Next we have to evaluate

$$\partial(\Delta_x - \Delta_y)\frac{1}{1+gD}\partial(\Delta_x - \Delta_y) =$$

$$= \int d^2z_1 d^2z_2\,\frac{\partial}{\partial z_1^\mu}\left(\Delta_F(x-z_1;\mu) - \Delta_F(y-z_1;\mu)\right)$$

$$\times\left[\frac{g^{\mu\nu}}{1+\zeta\frac{g}{\pi}}\delta^{(2)}(z_1-z_2) + \frac{g}{\pi}\frac{\chi}{\left(1+\zeta\frac{g}{\pi}\right)\left(1+(\zeta-\chi)\frac{g}{\pi}\right)}\frac{\partial}{\partial z_{1\mu}}\frac{\partial}{\partial z_{1\nu}}\Delta_F(z_1-z_2;\mu)\right]$$

$$\times\frac{\partial}{\partial z_2^\nu}\left(\Delta_F(x-z_2;\mu) - \Delta_F(y-z_2;\mu)\right)$$

$$
= \frac{1}{1+\zeta\frac{g}{\pi}} \int d^2z \frac{\partial}{\partial z^\mu}\Big(\Delta_F(x-z;\mu) - \Delta_F(y-z;\mu)\Big)\frac{\partial}{\partial z_\mu}\Big(\Delta_F(x-z;\mu) - \Delta_F(y-z;\mu)\Big)
$$

$$
+ \frac{g}{\pi}\frac{\chi}{\left(1+\zeta\frac{g}{\pi}\right)\left(1+(\zeta-\chi)\frac{g}{\pi}\right)} \int d^2z_1 d^2z_2 \frac{\partial}{\partial z_1^\mu}\Big(\Delta_F(x-z_1;\mu) - \Delta_F(y-z_1;\mu)\Big)
$$

$$
\times \frac{\partial}{\partial z_{1\mu}}\frac{\partial}{\partial z_{1\nu}}\Delta_F(z_1-z_2;\mu)\frac{\partial}{\partial z_2^\nu}\Big(\Delta_F(x-z_2;\mu) - \Delta_F(y-z_2;\mu)\Big)
$$

$$
= -\frac{1}{1+\zeta\frac{g}{\pi}} \int d^2z \Big(\Delta_F(x-z;\mu) - \Delta_F(y-z;\mu)\Big)\Big(\delta^{(2)}(x-z) - \delta^{(2)}(y-z)\Big)
$$

$$
- \frac{g}{\pi}\frac{\chi}{\left(1+\zeta\frac{g}{\pi}\right)\left(1+(\zeta-\chi)\frac{g}{\pi}\right)} \int d^2z_1 d^2z_2 \Big(\delta^{(2)}(x-z_1) - \delta^{(2)}(y-z_1)\Big)
$$

$$
\times \Delta_F(z_1-z_2)\Big(\delta^{(2)}(x-z_2) - \delta^{(2)}(y-z_2)\Big)
$$

$$
= -\frac{1}{1+\zeta\frac{g}{\pi}}\Big(2\Delta_F(0;\mu) - 2\Delta_F(x-y;\mu)\Big)
$$

$$
- \frac{g}{\pi}\frac{\chi}{\left(1+\zeta\frac{g}{\pi}\right)\left(1+(\zeta-\chi)\frac{g}{\pi}\right)}\Big(2\Delta_F(0;\mu) - 2\Delta_F(x-y;\mu)\Big)
$$

$$
= -\frac{2}{\left(1+\zeta\frac{g}{\pi}\right)\left(1+(\zeta-\chi)\frac{g}{\pi}\right)}\left(1+(\zeta-\chi)\frac{g}{\pi}+\frac{g}{\pi}\chi\right)\Big(\Delta_F(0;\mu) - \Delta_F(x-y;\mu)\Big)
$$

$$
= -\frac{2}{1+(\zeta-\chi)\frac{g}{\pi}}\Big(\Delta_F(0;\mu) - \Delta_F(x-y;\mu)\Big). \tag{A.80}
$$

Furthermore, we have to calculate

$$
\partial(\Delta_x - \Delta_y)\varepsilon\frac{1}{1+gD}\varepsilon\partial(\Delta_x - \Delta_y) =
$$

$$
= \int d^2z_1 d^2z_2 \frac{\partial}{\partial z_1^\mu}\Big(\Delta_F(x-z_1;\mu) - \Delta_F(y-z_1;\mu)\Big)\varepsilon^{\mu\nu}
$$

$$
\times \left[\frac{g_{\nu\rho}}{1+\zeta\frac{g}{\pi}}\delta^{(2)}(z_1-z_2) + \frac{g}{\pi}\frac{\chi}{\left(1+\zeta\frac{g}{\pi}\right)\left(1+(\zeta-\chi)\frac{g}{\pi}\right)}\frac{\partial}{\partial z_1^\nu}\frac{\partial}{\partial z_1^\rho}\Delta_F(z_1-z_2;\mu)\right]
$$

$$
\times \varepsilon^{\rho\sigma}\frac{\partial}{\partial z_2^\sigma}\Big(\Delta_F(x-z_2;\mu) - \Delta_F(y-z_2;\mu)\Big)
$$

$$
= -\frac{1}{1+\zeta\frac{g}{\pi}} \int d^2z \Big(\Delta_F(x-z;\mu) - \Delta_F(y-z;\mu)\Big)\Big(\delta^{(2)}(x-z) - \delta^{(2)}(y-z)\Big)
$$

$$
= -\frac{2}{1+\zeta\frac{g}{\pi}}\Big(\Delta_F(0;\mu) - \Delta_F(x-y;\mu)\Big), \tag{A.81}
$$

where we have used $\varepsilon^{\mu\nu}\varepsilon_{\nu\rho} = g^\mu_\rho$ and $\varepsilon^{\mu\nu}\partial_\mu\partial_\nu = 0$. Finally, the following expression has to be evaluated

$$
\partial(\Delta_x - \Delta_y)\varepsilon\frac{1}{1+gD}\partial(\Delta_x - \Delta_y) - \partial(\Delta_x - \Delta_y)\frac{1}{1+gD}\varepsilon\partial(\Delta_x - \Delta_y) =
$$

A.2. Two–point causal Green function $G(x, y)$

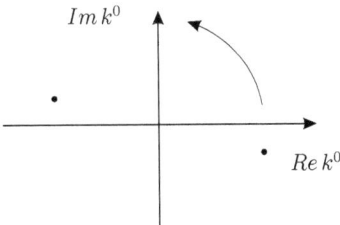

Figure A.2.: Since the poles of the causal Green function are as indicated in the figure the Wick rotation has to be performed counter-clockwise.

$$
\begin{aligned}
&= \int d^2z_1 d^2z_2 \, \frac{\partial}{\partial z_1^\mu} \Big(\Delta_F(x - z_1; \mu) - \Delta_F(y - z_1; \mu) \Big) \varepsilon^{\mu\nu} \\
&\quad \times \left[\frac{g_{\nu\rho}}{1 + \zeta \frac{g}{\pi}} \delta^{(2)}(z_1 - z_2) + \frac{g}{\pi} \frac{\chi}{\left(1 + \zeta \frac{g}{\pi}\right)\left(1 + (\zeta - \chi) \frac{g}{\pi}\right)} \frac{\partial}{\partial z_1^\nu} \frac{\partial}{\partial z_1^\rho} \Delta_F(z_1 - z_2; \mu) \right] \\
&\quad \times \frac{\partial}{\partial z_{2\rho}} \Big(\Delta_F(x - z_2; \mu) - \Delta_F(y - z_2; \mu) \Big) \\
&\quad - \int d^2z_1 d^2z_2 \, \frac{\partial}{\partial z_{1\nu}} \Big(\Delta_F(x - z_1; \mu) - \Delta_F(y - z_1; \mu) \Big) \\
&\quad \times \left[\frac{g_{\nu\rho}}{1 + \zeta \frac{g}{\pi}} \delta^{(2)}(z_1 - z_2) + \frac{g}{\pi} \frac{\chi}{\left(1 + \zeta \frac{g}{\pi}\right)\left(1 + (\zeta - \chi) \frac{g}{\pi}\right)} \frac{\partial}{\partial z_1^\nu} \frac{\partial}{\partial z_1^\rho} \Delta_F(z_1 - z_2; \mu) \right] \\
&\quad \times \varepsilon^{\rho\sigma} \frac{\partial}{\partial z_2^\sigma} \Big(\Delta_F(x - z_2; \mu) - \Delta_F(y - z_2; \mu) \Big) \\
&= 0, \quad\quad\quad\quad\quad\quad\quad\quad\quad\quad\quad\quad\quad\quad\quad\quad\quad\quad\quad\quad\quad\quad\quad\quad (A.82)
\end{aligned}
$$

where we have used again $\varepsilon^{\mu\nu} \partial_\mu \partial_\nu = 0$. The Green function reads

$$
\begin{aligned}
G(x, y) &= \frac{1}{2\pi} \frac{\gamma^\mu (x - y)_\mu}{(x - y)^2 - i0} \exp \Bigg\{ \frac{g}{1 + (\zeta - \chi)\frac{g}{\pi}} \Big(i\Delta_F(0; \mu) - i\Delta_F(x - y; \mu) \Big) \\
&\quad - \frac{g}{1 + \zeta \frac{g}{\pi}} \Big(i\Delta_F(0; \mu) - i\Delta_F(x - y; \mu) \Big) \Bigg\} \\
&= \frac{1}{2\pi} \frac{\gamma^\mu (x - y)_\mu}{(x - y)^2 - i0} \exp \Bigg\{ \frac{ig^2}{\pi} \frac{\chi}{\left(1 + (\zeta - \chi)\frac{g}{\pi}\right)\left(1 + \zeta \frac{g}{\pi}\right)} \Big(\Delta_F(0; \mu) - \Delta_F(x - y; \mu) \Big) \Bigg\}.
\end{aligned}
$$
(A.83)

A.2.2. Evaluation of $\Delta_F(0; \mu)$

Here we calculate $\Delta_F(0; \mu)$. Since for this causal Green function the singularities are as indicated in Fig. (A.2) we have to perform the Wick rotation counter-clockwise as indicated in the figure.

$$\begin{aligned}\Delta_F(0;\mu) &= \int \frac{d^2k}{(2\pi)^2} \frac{1}{-k^2+\mu^2-i0} \\ &= -\int \frac{dk_E dk^1}{(2\pi)^2 i} \frac{1}{k_E^2+k_1^2+\mu^2} \\ &= -\int_0^{\tilde{\Lambda}} \frac{dk}{2\pi i} \frac{k}{k^2+\mu^2} \\ &= -\int_{\mu^2}^{\Lambda^2} \frac{dl}{4\pi i} \frac{1}{l} \\ &= -\frac{1}{4\pi i} \ln\left(\frac{\Lambda^2}{\mu^2}\right),\end{aligned} \qquad (A.84)$$

where we have used Eq. (A.30), $k_E = -ik_0$, $k = k_E^2+k_1^2$ and $l = k^2+\mu^2$.

A.3. Two–point correlation function $C(x,y)$

Next we evaluate the two point correlation function

$$\begin{aligned}C(x,y) &= \text{Tr}\left\{\left(i\frac{\delta}{\delta J(x)}\right)\left(\frac{1-\gamma^5}{2}\right)\left(\frac{1}{i}\frac{\delta}{\delta \bar{J}(x)}\right)\right\} \\ &\quad \times \text{Tr}\left\{\left(i\frac{\delta}{\delta J(y)}\right)\left(\frac{1+\gamma^5}{2}\right)\left(\frac{1}{i}\frac{\delta}{\delta \bar{J}(y)}\right)\right\}Z_{\text{Th}}^{(g)}[A;J,\bar{J}]\bigg|_{A=J=\bar{J}=0} \\ &= \exp\left\{\frac{i}{2}g\int d^2z \frac{\delta}{\delta A_\mu(z)}\frac{\delta}{\delta A^\mu(z)}\right\} \\ &\quad \times \exp\left\{\frac{i}{2}\int d^2z_1 d^2z_2\, A^\mu(z_1)D^{(0)}_{\mu\nu}(z_1-z_2)A^\nu(z_2)\right\} \\ &\quad \times \text{Tr}\left\{\left(\frac{1-\gamma^5}{2}\right)S_A(x,y)\left(\frac{1+\gamma^5}{2}\right)S_A(y,x)\right\}\bigg|_{A=0} \\ &= \frac{1}{4\pi^2}\frac{(x-y)_\mu}{(x-y)^2-i0}\frac{(y-x)_\nu}{(y-x)^2-i0}\int \mathcal{D}^2u \exp\left\{-\frac{i}{2}\int d^2z\, u_\rho(z)u^\rho(z)\right. \\ &\quad \left. -\frac{i}{2}g\int d^2z_1 d^2z_2\, u^\rho(z_1)D^{(0)}_{\rho\sigma}(z_1-z_2)u^\sigma(z_2)\right\} \\ &\quad \times \text{Tr}\left\{\left(\frac{1-\gamma^5}{2}\right)\gamma^\mu \exp\left[-2\sqrt{g}\varepsilon^{\alpha\beta}\gamma^5\right.\right. \\ &\quad \left.\left. \times \int d^2z \frac{\partial}{\partial z^\alpha}[\Delta_F(x-z;\mu)-\Delta_F(y-z;\mu)]u_\beta(z)\right]\left(\frac{1+\gamma^5}{2}\right)\gamma^\nu\right\} \\ &= -\frac{1}{4\pi^2}\frac{1}{(x-y)^2-i0}\int \mathcal{D}^2u \exp\left\{-\frac{i}{2}\int d^2z\, u_\mu(z)u^\mu(z)\right. \\ &\quad -\frac{i}{2}g\int d^2z_1 d^2z_2\, u^\mu(z_1)D^{(0)}_{\mu\nu}(z_1-z_2)u^\nu(z_2) \\ &\quad \left. -2\sqrt{g}\varepsilon^{\alpha\beta}\int d^2z \frac{\partial}{\partial z^\alpha}[\Delta_F(x-z;\mu)-\Delta_F(y-z;\mu)]u_\beta(z)\right\},\end{aligned} \qquad (A.85)$$

where we have used Eq. (A.58), the relation

$$\text{Tr}\left\{\left(\frac{1-\gamma^5}{2}\right)\gamma^\mu e^{A\gamma^5}\left(\frac{1+\gamma^5}{2}\right)\gamma^\nu\right\} =$$

$$= \text{Tr}\left\{\gamma^\nu\left(\frac{1-\gamma^5}{2}\right)\gamma^\mu e^{A\gamma^5}\right\}$$
$$= \frac{1}{2}\text{Tr}\left\{\gamma^\nu(1-\gamma^5)\gamma^\mu\right\}\cosh A + \frac{1}{2}\text{Tr}\left\{\gamma^\nu(1-\gamma^5)\gamma^\mu\gamma^5\right\}\sinh A$$
$$= \frac{1}{2}\text{Tr}\left\{\gamma^\nu\gamma^\mu\right\}e^A + \frac{1}{2}\text{Tr}\left\{\gamma^\nu\gamma^\mu\gamma^5\right\}e^A$$
$$= g^{\nu\mu}e^A + \varepsilon^{\nu\mu}e^A, \tag{A.86}$$

and Eq. (A.6) in the last line. Using Eq. (A.57) with

$$L_{\mu\nu} = -\frac{i}{2}(1+gD)_{\mu\nu},$$
$$2\Delta^\mu L_{\mu\nu} = -2\sqrt{g}\,\partial(\Delta_x - \Delta_y)^\mu \varepsilon_{\mu\nu},$$
$$\Delta^\mu = -2i\sqrt{g}\,\partial(\Delta_x - \Delta_y)^\nu \varepsilon_{\nu\rho}((1+gD)^{-1})^{\rho\mu}, \tag{A.87}$$

and

$$-\Delta L \Delta = 2ig\partial(\Delta_x - \Delta_y)\varepsilon \frac{1}{1+gD}\varepsilon\partial(\Delta_x - \Delta_y), \tag{A.88}$$

we find

$$C(x,y) = -\frac{1}{4\pi^2}\frac{1}{(x-y)^2-i0}\exp\left\{2ig\partial(\Delta_x-\Delta_y)\varepsilon\frac{1}{1+gD}\varepsilon\partial(\Delta_x-\Delta_y)\right\}$$
$$= -\frac{1}{4\pi^2}\frac{1}{(x-y)^2-i0}\exp\left\{-\frac{4ig}{1+\zeta\frac{g}{\pi}}\left(\Delta_F(0;\mu)-\Delta_F(x-y;\mu)\right)\right\}. \tag{A.89}$$

A.4. The Schwinger term for the free theory

First we have to evaluate

$$\frac{\partial}{\partial x_0}\Delta_F(x-y;\mu) = \frac{\partial}{\partial x_0}\int\frac{d^2k}{(2\pi)^2}\frac{e^{-ik(x-y)}}{-k^2+\mu^2-i0}$$
$$= \int\frac{d^2k}{(2\pi)^2 i}\frac{k^0 e^{-ik(x-y)}}{-k^2+\mu^2-i0}$$
$$= \int\frac{dk^1}{2\pi}\frac{e^{ik^1(x^1-y^1)}}{2\bar{k}}\int\frac{dk^0}{2\pi i}k^0 e^{-ik^0(x^0-y^0)}\left(\frac{1}{k^0+\bar{k}}-\frac{1}{k^0-\bar{k}}\right)$$
$$= \int\frac{dk^1}{4\pi}e^{ik^1(x^1-y^1)}\left[\Theta(x^0-y^0)e^{-i\bar{k}(x^0-y^0)}-\Theta(y^0-x^0)e^{i\bar{k}(x^0-y^0)}\right], \tag{A.90}$$

with $\bar{k} = \sqrt{k_1^2+\mu^2-i0}$. The poles of the derivative of the causal function are as indicated in Fig. (A.1). The difference of the limits finally reads

$$\left(\lim_{x^0\to y^{0+}}-\lim_{x^0\to y^{0-}}\right)\frac{\partial}{\partial x_0}\frac{\partial}{\partial x_1}\Delta_F(x-y;\mu) = \frac{\partial}{\partial x_1}\delta(x^1-y^1). \tag{A.91}$$

A.5. The norm of the current states for the free theory

A.5.1. The time-ordered current correlation function

Using the current definition Eq. (A.33) we find for the time-ordered current correlation function for $x \neq y$

$$\langle 0|T(j^\mu(x)j^\nu(y))|0\rangle_{A=0} =$$
$$= \frac{1}{Z_{\text{Th}}^{(0)}[0; J, \bar{J}]} \text{Tr}\left\{\left(i\frac{\delta}{\delta J(x)}\right)\gamma^\mu\left(\frac{1}{i}\frac{\delta}{\delta \bar{J}(x)}\right)\right\}\text{Tr}\left\{\left(i\frac{\delta}{\delta J(y)}\right)\gamma^\nu\left(\frac{1}{i}\frac{\delta}{\delta \bar{J}(y)}\right)\right\}Z_{\text{Th}}^{(0)}[0; J, \bar{J}]\bigg|_{J=\bar{J}=0}$$
$$= \text{Tr}\left\{S_0(y,x)\gamma^\mu S_0(x,y)\gamma^\nu\right\}$$
$$= \frac{1}{4\pi^2}\frac{(y-x)_\alpha(x-y)_\beta}{(x-y)^4}\text{Tr}\left\{\gamma^\alpha\gamma^\mu\gamma^\beta\gamma^\nu\right\}$$
$$= \frac{1}{4\pi^2}\frac{(y-x)_\alpha(x-y)_\beta}{(x-y)^4}\text{Tr}\left(4g^{\alpha\mu}g^{\beta\nu} - 2g^{\alpha\beta}g^{\mu\nu}\right)$$
$$= \frac{1}{2\pi^2}\left(\frac{g^{\mu\nu}}{(x-y)^2} - \frac{2(x-y)^\mu(x-y)^\nu}{(x-y)^4}\right)$$
$$= \frac{i}{\pi}\partial^\mu\partial^\nu\Delta_F(x-y;\mu), \tag{A.92}$$

where we have used the free Green function Eq. (A.31), the relation

$$\text{Tr}\left\{\gamma^\alpha\gamma^\mu\gamma^\beta\gamma^\nu\right\} = \text{Tr}\left\{(g^{\alpha\mu} + \varepsilon^{\alpha\mu}\gamma^5)(g^{\beta\nu} + \varepsilon^{\beta\nu}\gamma^5)\right\}$$
$$= 2g^{\alpha\mu}g^{\beta\nu} + 2\varepsilon^{\alpha\mu}\varepsilon^{\beta\nu}$$
$$= 2g^{\alpha\mu}g^{\beta\nu} + 2g^{\alpha\nu}g^{\mu\beta} - 2g^{\alpha\beta}g^{\mu\nu}, \tag{A.93}$$

and the causal Green function Eq. (A.30). For $x = y$ one has to take spatial point splitting regularisation into account, which yields the additional non-covariant part containing the delta function (see Eq. (1.56)).

A.5.2. The norm of the states

Using the Wightman function

$$D^{(\pm)}(x-y) = \int \frac{d^2k}{(2\pi)^2} 2\pi\theta(k^0)\delta(k^2) e^{\mp ik\cdot(x-y)}$$
$$= \frac{1}{4\pi}\int \frac{dk^1}{\sqrt{k_1^2+\mu^2}} e^{\mp i\sqrt{k_1^2+\mu^2}(x^0-y^0) \pm ik^1(x^1-y^1)}, \tag{A.94}$$

we find the following relation

$$\Delta_F(x-y;\mu) = i\theta(x^0-y^0) D^{(+)}(x-y) + i\theta(y^0-x^0) D^{(-)}(x-y). \tag{A.95}$$

A.5. The norm of the current states for the free theory

Furthermore, we have

$$\langle 0|T(j^\mu(x)j^\nu(y))|0\rangle = \theta(x^0 - y^0)\langle 0|j^\mu(x)j^\nu(y)|0\rangle + \theta(y^0 - x^0)\langle 0|j^\nu(y)j^\mu(x)|0\rangle. \quad (A.96)$$

Now we will prove that the current operator product reads

$$\langle 0|j^\mu(x)j^\nu(y)|0\rangle = -\frac{1}{\pi}\frac{\partial}{\partial x_\mu}\frac{\partial}{\partial x_\nu}D^{(+)}(x-y). \quad (A.97)$$

We have

$$\theta(x^0 - y^0)\langle 0|j^\mu(x)j^\nu(y)|0\rangle + \theta(y^0 - x^0)\langle 0|j^\nu(y)j^\mu(x))|0\rangle =$$

$$= -\theta(x^0 - y^0)\frac{1}{\pi}\partial^\mu\partial^\nu D^{(+)}(x-y) - \theta(y^0 - x^0)\frac{1}{\pi}\partial^\mu\partial^\nu D^{(+)}(y-x)$$

$$= \frac{i}{\pi}\partial^\mu\partial^\nu\Delta_F(x-y;\mu) + \frac{1}{\pi}(g^{\alpha\mu}g^{\beta\nu} + g^{\alpha\nu}g^{\beta\mu})$$

$$\times \left(\partial_\alpha\Theta(x^0 - y^0)\partial_\beta D^{(+)}(x-y) + \partial_\alpha\Theta(y^0 - x^0)\partial_\beta D^{(+)}(y-x)\right)$$

$$+ \frac{1}{\pi}\left(\partial_\mu\partial_\nu\Theta(x^0 - y^0)D^{(+)}(x-y) + \partial_\mu\partial_\nu\Theta(y^0 - x^0)D^{(+)}(y-x)\right)$$

$$= \frac{i}{\pi}\partial^\mu\partial^\nu\Delta_F(x-y;\mu) + \frac{1}{\pi}(g^{0\mu}g^{\beta\nu} + g^{0\nu}g^{\beta\mu})$$

$$\times \left(\delta(x^0 - y^0)\partial_\beta D^{(+)}(x-y) - \delta(y^0 - x^0)\partial_\beta D^{(+)}(y-x)\right)$$

$$+ \frac{1}{\pi}g^{0\mu}g^{0\nu}\left(\partial_0\delta(x^0 - y^0)D^{(+)}(x-y) - \partial_0\delta(y^0 - x^0)D^{(+)}(y-x)\right)$$

$$= \frac{i}{\pi}\partial^\mu\partial^\nu\Delta_F(x-y;\mu) + \frac{2}{\pi}g^{0\mu}g^{0\nu}\delta(x^0 - y^0)\partial_0\left(D^{(+)}(x-y) - D^{(+)}(y-x)\right)$$

$$- \frac{1}{\pi}g^{0\mu}g^{0\nu}\delta(x^0 - y^0)\partial_0\left(D^{(+)}(x-y) - D^{(+)}(y-x)\right)$$

$$= \frac{i}{\pi}\partial^\mu\partial^\nu\Delta_F(x-y;\mu) - \frac{i}{\pi}g^{0\mu}g^{0\nu}\delta^{(2)}(x-y)$$

$$= \langle 0|T(j^\mu(x)j^\nu(y))|0\rangle, \quad (A.98)$$

where we have used

$$\delta(x^0 - y^0)\left(D^{(+)}(x-y) - D^{(+)}(y-x)\right) = 0, \quad (A.99)$$

and

$$\delta(x^0 - y^0)\partial_0\left(D^{(+)}(x-y) - D^{(+)}(y-x)\right) = -i\delta^{(2)}(x-y), \quad (A.100)$$

and all derivatives are understood with respect to x.

B. Sine–Gordon Model

B.1. Expansion to second order in α

B.1.1. The massless sine–Gordon model

The second order correction to the two–point Green function in α–expansion is defined by

$$\begin{aligned}
G^{(2)}(x,y) &= -\frac{1}{2}\iint d^2z_1 d^2z_2 \,\langle 0|\mathrm{T}(\varphi(x)\varphi(y)\mathcal{L}_i(z_1)\mathcal{L}_i(z_2))|0\rangle_c \\
&= -\frac{1}{2}\alpha^2 \sum_{m_1=2}^{\infty}\frac{(-1)^{m_1}}{(2m_1)!}\beta^{2(m_1-1)} \sum_{m_2=2}^{\infty}\frac{(-1)^{m_2}}{(2m_2)!}\beta^{2(m_2-1)} \\
&\quad \times \iint d^2z_1 d^2z_2 \,\langle 0|\mathrm{T}(\varphi(x)\varphi(y)\varphi^{2m_1}(z_1)\varphi^{2m_2}(z_2))|0\rangle_c \\
&\quad - \alpha^2(Z_1-1) \sum_{m_1=2}^{\infty}\frac{(-1)^{m_1}}{(2m_1)!}\beta^{2(m_1-1)} \sum_{m_2=1}^{\infty}\frac{(-1)^{m_2}}{(2m_2)!}\beta^{2(m_2-1)} \\
&\quad \times \iint d^2z_1 d^2z_2 \,\langle 0|\mathrm{T}(\varphi(x)\varphi(y)\varphi^{2m_1}(z_1)\varphi^{2m_2}(z_2))|0\rangle_c \\
&\quad - \frac{1}{2}\alpha^2(Z_1-1)^2 \sum_{m_1=1}^{\infty}\frac{(-1)^{m_1}}{(2m_1)!}\beta^{2(m_1-1)} \sum_{m_2=1}^{\infty}\frac{(-1)^{m_2}}{(2m_2)!}\beta^{2(m_2-1)} \\
&\quad \times \iint d^2z_1 d^2z_2 \,\langle 0|\mathrm{T}(\varphi(x)\varphi(y)\varphi^{2m_1}(z_1)\varphi^{2m_2}(z_2))|0\rangle_c .
\end{aligned} \quad (B.1)$$

The time-ordered product equals the sum of all complete contractions. The only non-vanishing contributions correspond to the Feynman diagrams of Fig. B.1 and Fig. B.2. Therefore, we evaluate the "multiplicity" M (the number how often the same diagram is reproduced by all complete contractions of the time-ordered product) of both diagrams. We suppose that there are $2m_1$ fields at z_1 and $2m_2$ fields at z_2. In Fig. B.1 there are $2m$ propagators connecting z_1 with z_2, while $2m-1$ propagators connect z_1 with z_2 in Fig. B.2. The multiplicity M_1 of the diagram in Fig. B.1 is obtained by the following consideration. The first external propagator can be connected $2m_1$ times at z_2 while the second external propagator can be connected $2m_1 - 1$ times (I). The $2m$ propagators linking z_2 with z_2 have $2m_1 - 2$, $2m_1 - 3$ and so on possibilities at z_2, $2m_2$, $2m_2 - 1$ and so on at z_2 – but now we have over-counted since interchanging them yields the same contraction. Therefore, we have to divide by $(2m)!$ (II). The "beginning" of the first "loop" propagator seated at z_2 has $2m_1 - 2m - 2$ possibilities to attach while for his "ending" there are only $2m_1 - 2m - 3$ possibilities left and so on. There are $m_1 - m - 1$ such closed loops at z_2. Since interchanging "beginning" with "ending" of each "loop" propagator yields the same contraction, we have over-counted and have to divide by 2^{m_1-m-1}. Since interchanging these propagators among themselves yields the same contraction we have to divide by $(m_1 - m - 1)!$ (III). The "beginning" of the first

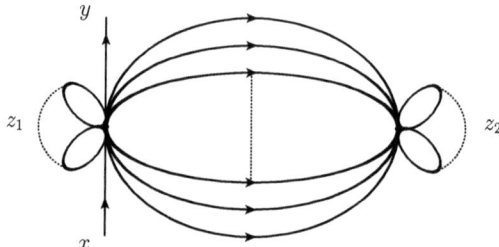

Figure B.1.: Feynman diagram for corrections to the two–point Green function to second order in α and to arbitrary order in β^2. This diagram corresponds to a non–vanishing expectation value for an even number of internal lines between the two vertices only. The number of tadpoles at the vertex z_1 equals $n_1 = m_1 - m - 1$, while the corresponding number at z_2 is $n_2 = m_2 - m$. The number of propagators linking z_1 with z_2 is $2m$.

"loop" propagator seated at z_2 has $2m_2 - 2m$ possibilities to attach while for his "ending" there are only $2m_2 - 2m - 1$ possibilities left and so on. There are $m_2 - m$ such closed loops at z_2. Due to over-counting we divide by 2^{m_2-m} (-interchanging "beginning" with "ending" of each "loop" propagator yields the same contraction) and with $(m_2 - m)!$ (-interchanging these propagators among themselves yields the same contraction) (IV). Thus we are lead to

$$M_1 = \underbrace{(2m_1)(2m_1 - 1)}_{I} \underbrace{\frac{(2m_1 - 2)(2m_1 - 3)\cdots(2m_1 - 1 - 2m)(2m_2)\cdots(2m_2 + 1 - 2m)}{(2m)!}}_{II}$$

$$\times \underbrace{\frac{(2m_1 - 2 - 2m)\cdots 1}{(m_1 - m - 1)!2^{m_1 - m - 1}}}_{III} \underbrace{\frac{(2m_2 - 2m)\cdots 1}{(m_2 - m)!2^{m_2 - m}}}_{IV}$$

$$= \frac{(2m_1)!(2m_2)!}{(2m)!(m_1 - m - 1)!(m_2 - m)!2^{m_1 + m_2 - 2m - 1}}. \tag{B.2}$$

Analogous reasoning yields for Fig. B.2

$$M_2 = (2m_1)(2m_2)\frac{(2m_1 - 1)(2m_1 - 2)\cdots(2m_1 + 1 - 2m)(2m_2 - 1)\cdots(2m_2 + 1 - 2m)}{(2m - 1)!}$$

$$\times \frac{(2m_1 - 2m)\cdots 1}{(m_1 - m)!2^{m_1 - m}}\frac{(2m_2 - 2m)\cdots 1}{(m_2 - m)!2^{m_2 - m}}$$

$$= \frac{(2m_1)!(2m_2)!}{(2m - 1)!(m_1 - m)!(m_2 - m)!2^{m_1 + m_2 - 2m}}. \tag{B.3}$$

Next we evaluate

$$\sum_{m_1=A}^{\infty} \frac{(-1)^{m_1}}{(2m_1)!} \beta^{2(m_1-1)} \sum_{m_2=B}^{\infty} \frac{(-1)^{m_2}}{(2m_2)!} \beta^{2(m_2-1)} \iint d^2z_1 d^2z_2 \, \langle 0|T(\varphi(x)\varphi(y)\varphi^{2m_1}(z_1)\varphi^{2m_2}(z_2))|0\rangle_c =$$

$$= \sum_{m_1=A}^{\infty} \sum_{m_2=B}^{\infty} \frac{(-1)^{m_1+m_2}\beta^{2(m_1+m_2-2)}}{(2m_1)!(2m_2)!}$$

B.1. Expansion to second order in α

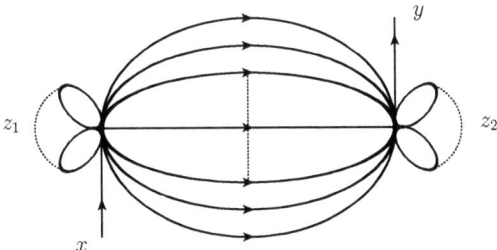

Figure B.2.: Feynman diagram for corrections to the two–point Green function to second order in α and to arbitrary order in β^2. This diagram corresponds to a non–vanishing expectation value for an odd number of internal lines between two vertices. The number of tadpoles for this diagram at z_1 equals $n_1^* = m_1 - m$ and for the vertex at z_2 we find $n_2 = m_2 - m$. The number of propagators linking z_1 with z_2 is $2m - 1$ in this case.

$$\times \left\{ (1 - \delta_{m_1 1}) \sum_{m=1}^{N_1} \frac{(2m_1)!(2m_2)!}{(2m)!(m_1 - m - 1)!(m_2 - m)!2^{m_1+m_2-2m-1}} \right.$$

$$\times \iint d^2 z_1 d^2 z_2 \left([-i\Delta_F(x - z_1)][-i\Delta_F(y - z_1)][-i\Delta_F(z_1 - z_2)]^{2m}[-i\Delta_F(0)]^{m_1+m_2-2m-1} \right.$$

$$\left. + (z_1 \leftrightarrow z_2, m_1 \leftrightarrow m_2) \right)$$

$$+ \sum_{m=1}^{N_2} \frac{(2m_1)!(2m_2)!}{(2m-1)!(m_1 - m)!(m_2 - m)!2^{m_1+m_2-2m}}$$

$$\times \iint d^2 z_1 d^2 z_2 \left([-i\Delta_F(x - z_1)][-i\Delta_F(y - z_2)][-i\Delta_F(z_1 - z_2)]^{2m-1}[-i\Delta_F(0)]^{m_1+m_2-2m} \right.$$

$$\left. \left. + (z_1 \leftrightarrow z_2, m_1 \leftrightarrow m_2) \right) \right\}$$

$$= \sum_{m=1}^{\infty} \sum_{n_1=0}^{\infty} \sum_{n_2=\delta_{m_1}\delta_{B_2}}^{\infty} \frac{(-1)^{n_1+n_2+2m+1} \beta^{2(n_1+n_2+2m-1)}}{(2m)!n_1!n_2!2^{n_1+n_2}}$$

$$\times \iint d^2 z_1 d^2 z_2 [-i\Delta_F(x - z_1)][-i\Delta_F(y - z_1)][-i\Delta_F(z_1 - z_2)]^{2m}[-i\Delta_F(0)]^{n_1+n_2}$$

$$+ (A \leftrightarrow B)$$

$$+ \sum_{m=1}^{\infty} \sum_{n_1^*=\delta_{m_1}\delta_{A_2}}^{\infty} \sum_{n_2=\delta_{m_1}\delta_{B_2}}^{\infty} \frac{(-1)^{n_1^*+n_2+2m} \beta^{2(n_1^*+n_2+2m-2)}}{(2m-1)!n_1^*!n_2!2^{n_1^*+n_2}}$$

$$\times \iint d^2 z_1 d^2 z_2 [-i\Delta_F(x - z_1)][-i\Delta_F(y - z_2)][-i\Delta_F(z_1 - z_2)]^{2m-1}[-i\Delta_F(0)]^{n_1^*+n_2}$$

$$+ (A \leftrightarrow B)$$

$$= - \sum_{m=1}^{\infty} \frac{\beta^{4m-2}}{(2m)!} \exp\left\{ i\frac{\beta^2}{2}\Delta_F(0) \right\} \left(\exp\left\{ i\frac{\beta^2}{2}\Delta_F(0) \right\} - \delta_{m_1}\delta_{B_2} \right)$$

$$\times \iint d^2 z_1 d^2 z_2 [-i\Delta_F(x - z_1)][-i\Delta_F(y - z_1)][-i\Delta_F(z_1 - z_2)]^{2m}$$

$$- (A \leftrightarrow B)$$

$$+ \sum_{m=1}^{\infty} \frac{\beta^{4m-4}}{(2m-1)!} \left(\exp\left\{ i\frac{\beta^2}{2}\Delta_F(0) \right\} - \delta_{m_1}\delta_{A_2} \right) \left(\exp\left\{ i\frac{\beta^2}{2}\Delta_F(0) \right\} - \delta_{m_1}\delta_{B_2} \right)$$

$$\times \iint d^2z_1 d^2z_2 [-i\Delta_F(x-z_1)][-i\Delta_F(y-z_2)][-i\Delta_F(z_1-z_2)]^{2m-1}$$
$$+ (A \leftrightarrow B)$$
$$= -\sum_{m=1}^{\infty} \frac{\beta^{4m-2}}{(2m)!} \exp\left\{i\frac{\beta^2}{2}\Delta_F(0)\right\} \left(2\exp\left\{i\frac{\beta^2}{2}\Delta_F(0)\right\} - \delta_{m1}\delta_{A2} - \delta_{m1}\delta_{B2}\right)$$
$$\times \iint d^2z_1 d^2z_2 [-i\Delta_F(x-z_1)][-i\Delta_F(y-z_1)][-i\Delta_F(z_1-z_2)]^{2m}$$
$$+ 2\sum_{m=1}^{\infty} \frac{\beta^{4m-4}}{(2m-1)!} \left(\exp\left\{i\frac{\beta^2}{2}\Delta_F(0)\right\} - \delta_{m1}\delta_{A2}\right)\left(\exp\left\{i\frac{\beta^2}{2}\Delta_F(0)\right\} - \delta_{m1}\delta_{B2}\right)$$
$$\times \iint d^2z_1 d^2z_2 [-i\Delta_F(x-z_1)][-i\Delta_F(y-z_2)][-i\Delta_F(z_1-z_2)]^{2m-1}, \quad (B.4)$$

with $n_1 = m_1 - m - 1$ being the number of closed loops at z_1 in Fig. B.1, $n_1^* = m_1 - m$ the corresponding number in Fig. B.2, $n_2 = m_2 - m$ the number of closed loops at z_2 in Fig. B.1 and Fig. B.2. The variables A and B take the values 1, 2 and $N_1 = \Theta(m_1 - 1 - m_2)m_2 + \Theta(m_2 - m_1 + 1)(m_1 - 1)$, $N_2 = \Theta(m_1 - m_2)m_2 + \Theta(m_2 - m_1)m_1$. Hence we obtain

$$G^{(2)}(x,y) = \frac{1}{2}\alpha^2 \sum_{m=1}^{\infty} \frac{\beta^{4m-2}}{(2m)!} \exp\left\{i\frac{\beta^2}{2}\Delta_F(0)\right\}\left(2\exp\left\{i\frac{\beta^2}{2}\Delta_F(0)\right\} - 2\delta_{m1}\right)$$
$$\times \iint d^2z_1 d^2z_2 [-i\Delta_F(x-z_1)][-i\Delta_F(y-z_1)][-i\Delta_F(z_1-z_2)]^{2m}$$
$$- \alpha^2 \sum_{m=1}^{\infty} \frac{\beta^{4m-4}}{(2m-1)!}\left(\exp\left\{i\frac{\beta^2}{2}\Delta_F(0)\right\} - \delta_{m1}\right)\left(\exp\left\{i\frac{\beta^2}{2}\Delta_F(0)\right\} - \delta_{m1}\right)$$
$$\times \iint d^2z_1 d^2z_2 [-i\Delta_F(x-z_1)][-i\Delta_F(y-z_2)][-i\Delta_F(z_1-z_2)]^{2m-1}$$
$$+ \alpha^2(Z_1 - 1)\sum_{m=1}^{\infty} \frac{\beta^{4m-2}}{(2m)!} \exp\left\{i\frac{\beta^2}{2}\Delta_F(0)\right\}\left(2\exp\left\{i\frac{\beta^2}{2}\Delta_F(0)\right\} - \delta_{m1}\right)$$
$$\times \iint d^2z_1 d^2z_2 [-i\Delta_F(x-z_1)][-i\Delta_F(y-z_1)][-i\Delta_F(z_1-z_2)]^{2m}$$
$$- 2\alpha^2(Z_1 - 1)\sum_{m=1}^{\infty} \frac{\beta^{4m-4}}{(2m-1)!}\left(\exp\left\{i\frac{\beta^2}{2}\Delta_F(0)\right\} - \delta_{m1}\right)\exp\left\{i\frac{\beta^2}{2}\Delta_F(0)\right\}$$
$$\times \iint d^2z_1 d^2z_2 [-i\Delta_F(x-z_1)][-i\Delta_F(y-z_2)][-i\Delta_F(z_1-z_2)]^{2m-1}$$
$$+ \alpha^2(Z_1 - 1)^2 \sum_{m=1}^{\infty} \frac{\beta^{4m-2}}{(2m)!} \exp\left\{i\frac{\beta^2}{2}\Delta_F(0)\right\}^2$$
$$\times \iint d^2z_1 d^2z_2 [-i\Delta_F(x-z_1)][-i\Delta_F(y-z_1)][-i\Delta_F(z_1-z_2)]^{2m}$$
$$- \alpha^2(Z_1 - 1)^2 \sum_{m=1}^{\infty} \frac{\beta^{4m-4}}{(2m-1)!} \exp\left\{i\frac{\beta^2}{2}\Delta_F(0)\right\}^2$$
$$\times \iint d^2z_1 d^2z_2 [-i\Delta_F(x-z_1)][-i\Delta_F(y-z_2)][-i\Delta_F(z_1-z_2)]^{2m-1}$$
$$= -\alpha^2 Z_1 \frac{\beta^2}{2} \exp\left\{i\frac{\beta^2}{2}\Delta_F(0)\right\}$$
$$\times \iint d^2z_1 d^2z_2 [-i\Delta_F(x-z_1)][-i\Delta_F(y-z_1)][-i\Delta_F(z_1-z_2)]^2$$
$$- \alpha^2\left(1 - Z_1 \exp\left\{i\frac{\beta^2}{2}\Delta_F(0)\right\}\right)^2$$

B.1. Expansion to second order in α

$$\times \iint d^2z_1 d^2z_2 [-i\Delta_F(x-z_1)][-i\Delta_F(y-z_2)][-i\Delta_F(z_1-z_2)]$$

$$+ Z_1^2 \alpha^2 \sum_{m=1}^{\infty} \frac{\beta^{4m-2}}{(2m)!} \exp\left\{i\frac{\beta^2}{2}\Delta_F(0)\right\}^2$$

$$\times \iint d^2z_1 d^2z_2 [-i\Delta_F(x-z_1)][-i\Delta_F(y-z_1)][-i\Delta_F(z_1-z_2)]^{2m}$$

$$- Z_1^2 \alpha^2 \sum_{m=2}^{\infty} \frac{\beta^{4m-4}}{(2m-1)!} \exp\left\{i\frac{\beta^2}{2}\Delta_F(0)\right\}^2$$

$$\times \iint d^2z_1 d^2z_2 [-i\Delta_F(x-z_1)][-i\Delta_F(y-z_2)][-i\Delta_F(z_1-z_2)]^{2m-1}$$

$$= -\alpha^2 Z_1 \frac{\beta^2}{2} \exp\left\{i\frac{\beta^2}{2}\Delta_F(0)\right\}$$

$$\times \iint d^2z_1 d^2z_2 [-i\Delta_F(x-z_1)][-i\Delta_F(y-z_1)][-i\Delta_F(z_1-z_2)]^2$$

$$- \alpha^2 \left(1 - Z_1 \exp\left\{i\frac{\beta^2}{2}\Delta_F(0)\right\}\right)^2$$

$$\times \iint d^2z_1 d^2z_2 [-i\Delta_F(x-z_1)][-i\Delta_F(y-z_2)][-i\Delta_F(z_1-z_2)]$$

$$+ \frac{Z_1^2 \alpha^2}{\beta^2} \exp\left\{i\beta^2 \Delta_F(0)\right\}$$

$$\times \iint d^2z_1 d^2z_2 [-i\Delta_F(x-z_1)]\left(\cosh[(-i\beta^2 \Delta_F(z_1-z_2)] - 1\right)[-i\Delta_F(y-z_1)]$$

$$- \frac{Z_1^2 \alpha^2}{\beta^2} \exp\left\{i\beta^2 \Delta_F(0)\right\}$$

$$\times \iint d^2z_1 d^2z_2 [-i\Delta_F(x-z_1)]\left(\sinh[(-i\beta^2 \Delta_F(z_1-z_2)] + i\beta^2 \Delta_F(z_1-z_2)\right)$$

$$\times [-i\Delta_F(y-z_2)]. \tag{B.5}$$

B.1.2. Momentum representation

Next we evaluate

$$\iint d^2z_1 d^2z_2 [-i\Delta_F(x-z_1)][-i\Delta_F(y-z_1)][-i\Delta_F(z_1-z_2)]^{2m} =$$

$$= \iint d^2z_1 d^2z_2 \left(\int \frac{d^2k}{(2\pi)^2 i} \frac{e^{-ik(x-z_1)}}{\alpha - k^2 - i0}\right)\left(\int \frac{d^2k}{(2\pi)^2 i} \frac{e^{-ik(y-z_1)}}{\alpha - k^2 - i0}\right)\left(\int \frac{d^2k}{(2\pi)^2 i} \frac{e^{-ik(z_1-z_2)}}{\alpha - k^2 - i0}\right)^{2m}$$

$$= \int \frac{d^2k_1}{(2\pi)^2} \frac{-i}{\alpha - k_1^2 - i0} \cdots \int \frac{d^2k_{2m+2}}{(2\pi)^2} \frac{-i}{\alpha - k_{2m+2}^2 - i0}$$

$$\times \iint d^2z_1 d^2z_2\, e^{-ik_1 x - ik_2 y - iz_1(k_3 + \cdots + k_{2m+2} - k_1 - k_2) + iz_2(k_3 + \cdots + k_{2m+2})}$$

$$= \int \frac{d^2k_1}{(2\pi)^2} \frac{-i}{\alpha - k_1^2 - i0} \cdots \int \frac{d^2k_{2m+2}}{(2\pi)^2} \frac{-i}{\alpha - k_{2m+2}^2 - i0}$$

$$\times e^{-ik_1 x - ik_2 y} \delta^{(2)}(k_3 + \cdots + k_{2m+2} - k_1 - k_2)\delta^{(2)}(k_3 + \cdots + k_{2m+2})(2\pi)^4$$

$$= \int \frac{d^2k_2}{(2\pi)^2} e^{ik_2(x-y)} \left(\frac{-i}{\alpha - k_2^2 - i0}\right)^2 \int \frac{d^2k_3}{(2\pi)^2} \frac{-i}{\alpha - k_3^2 - i0} \cdots$$

$$\times \int \frac{d^2 k_{2m}}{(2\pi)^2} \frac{-i}{\alpha - k_{2m}^2 - i0} \int \frac{d^2 k_{2m+1}}{(2\pi)^2} \frac{-i}{\alpha - k_{2m+1}^2 - i0} \frac{-i}{\alpha - (k_3 + \cdots + k_{2m+1})^2 - i0}. \quad \text{(B.6)}$$

Hence the momentum-representation is given by

$$\int d^2x\, e^{ipx} \iint d^2z_1 d^2z_2 [-i\Delta_F(x-z_1)][-i\Delta_F(z_1)][-i\Delta_F(z_1-z_2)]^{2m} =$$

$$= \int d^2x\, e^{ipx} \int \frac{d^2k_2}{(2\pi)^2} e^{ik_2 x} \left(\frac{-i}{\alpha - k_2^2 - i0}\right)^2 \int \frac{d^2k_3}{(2\pi)^2} \frac{-i}{\alpha - k_3^2 - i0} \cdots$$

$$\times \int \frac{d^2 k_{2m}}{(2\pi)^2} \frac{-i}{\alpha - k_{2m}^2 - i0} \int \frac{d^2 k_{2m+1}}{(2\pi)^2} \frac{-i}{\alpha - k_{2m+1}^2 - i0} \frac{-i}{\alpha - (k_3 + \cdots + k_{2m+1})^2 - i0}$$

$$= \left(\frac{-i}{\alpha - p^2 - i0}\right)^2 \int \frac{d^2k_3}{(2\pi)^2} \frac{-i}{\alpha - k_3^2 - i0} \cdots$$

$$\times \int \frac{d^2 k_{2m}}{(2\pi)^2} \frac{-i}{\alpha - k_{2m}^2 - i0} \int \frac{d^2 k_{2m+1}}{(2\pi)^2} \frac{-i}{\alpha - k_{2m+1}^2 - i0} \frac{-i}{\alpha - (k_3 + \cdots + k_{2m+1})^2 - i0}, \quad \text{(B.7)}$$

respectively

$$\int d^2x\, e^{ipx} \iint d^2z_1 d^2z_2 [-i\Delta_F(x-z_1)][-i\Delta_F(z_2)][-i\Delta_F(z_1-z_2)]^{2m-1} =$$

$$= \iint d^2z_1 d^2z_2 \left(\int \frac{d^2k}{(2\pi)^2 i} \frac{e^{-ik(x-z_1)}}{\alpha - k^2 - i0}\right) \left(\int \frac{d^2k}{(2\pi)^2 i} \frac{e^{-ikz_2}}{\alpha - k^2 - i0}\right) \left(\int \frac{d^2k}{(2\pi)^2 i} \frac{e^{-ik(z_1-z_2)}}{\alpha - k^2 - i0}\right)^{2m-1}$$

$$= \int \frac{d^2k_1}{(2\pi)^2} \frac{-i}{\alpha - k_1^2 - i0} \cdots \int \frac{d^2 k_{2m+1}}{(2\pi)^2} \frac{-i}{\alpha - k_{2m+1}^2 - i0}$$

$$\times \iint d^2z_1 d^2z_2\, e^{-ik_1 x - iz_1(k_3 + \cdots + k_{2m+1} - k_1) + iz_2(k_3 + \cdots + k_{2m+1} - k_2)}$$

$$= \int \frac{d^2k_1}{(2\pi)^2} \frac{-i}{\alpha - k_1^2 - i0} \cdots \int \frac{d^2 k_{2m+1}}{(2\pi)^2} \frac{-i}{\alpha - k_{2m+1}^2 - i0}$$

$$\times e^{-ik_1 x} \delta^{(2)}(k_3 + \cdots + k_{2m+1} - k_1) \delta^{(2)}(k_3 + \cdots + k_{2m+1} - k_2)(2\pi)^4$$

$$= \int \frac{d^2k_3}{(2\pi)^2} \frac{-i}{\alpha - k_3^2 - i0} \cdots \int \frac{d^2 k_{2m+1}}{(2\pi)^2} \frac{-i}{\alpha - k_{2m+1}^2 - i0}$$

$$\times e^{-i(k_3 + \cdots + k_{2m+1})x} \left(\frac{-i}{\alpha - (k_3 + \cdots + k_{2m+1})^2 - i0}\right)^2, \quad \text{(B.8)}$$

and

$$\int d^2x\, e^{ipx} \iint d^2z_1 d^2z_2 [-i\Delta_F(x-z_1)][-i\Delta_F(z_2)][-i\Delta_F(z_1-z_2)]^{2m-1} =$$

$$= \int d^2x\, e^{ipx} \int \frac{d^2k_3}{(2\pi)^2} \frac{-i}{\alpha - k_3^2 - i0} \cdots \int \frac{d^2 k_{2m+1}}{(2\pi)^2} \frac{-i}{\alpha - k_{2m+1}^2 - i0}$$

$$\times e^{-i(k_3 + \cdots + k_{2m+1})x} \left(\frac{-i}{\alpha - (k_3 + \cdots + k_{2m+1})^2 - i0}\right)^2$$

$$= \left(\frac{-i}{\alpha - p^2 - i0}\right)^2 \int \frac{d^2k_3}{(2\pi)^2} \frac{-i}{\alpha - k_3^2 - i0} \cdots \int \frac{d^2 k_{2m}}{(2\pi)^2} \frac{-i}{\alpha - k_{2m}^2 - i0} \frac{-i}{\alpha - (p - k_3 - \cdots k_{2m})^2 - i0}.$$

$$\text{(B.9)}$$

B.1. Expansion to second order in α

We have to deal with the special case $m = 1$ from above separately. We find

$$\iint d^2z_1 d^2z_2 [-i\Delta_F(x-z_1)][-i\Delta_F(z_2)][-i\Delta_F(z_1-z_2)] = \int \frac{d^2k}{(2\pi)^2} e^{-ikx} \left(\frac{-i}{\alpha - k^2 - i0}\right)^3,$$

(B.10)

and

$$\int d^2x\, e^{ipx} \iint d^2z_1 d^2z_2 [-i\Delta_F(x-z_1)][-i\Delta_F(z_2)][-i\Delta_F(z_1-z_2)] =$$

$$= \int d^2x\, e^{ipx} \int \frac{d^2k}{(2\pi)^2} e^{-ikx} \left(\frac{-i}{\alpha - k^2 - i0}\right)^3$$

$$= \left(\frac{-i}{\alpha - p^2 - i0}\right)^3.$$

(B.11)

Hence we obtain for the Green function in momentum-representation

$$\tilde{G}^{(2)}(p) = -\alpha^2 Z_1 \frac{\beta^2}{2} \exp\left\{i\frac{\beta^2}{2}\Delta_F(0)\right\} \left(\frac{-i}{\alpha - p^2 - i0}\right)^2 \int \frac{d^2k}{(2\pi)^2} \left(\frac{-i}{\alpha - k^2 - i0}\right)^2$$

$$- \alpha^2 \left(1 - Z_1 \exp\left\{i\frac{\beta^2}{2}\Delta_F(0)\right\}\right)^2 \left(\frac{-i}{\alpha - p^2 - i0}\right)^3$$

$$+ Z_1^2 \alpha^2 \sum_{m=1}^{\infty} \frac{\beta^{4m-2}}{(2m)!} \exp\left\{i\beta^2 \Delta_F(0)\right\}$$

$$\times \left(\frac{-i}{\alpha - p^2 - i0}\right)^2 \int \frac{d^2k_1}{(2\pi)^2} \frac{-i}{\alpha - k_1^2 - i0} \cdots \int \frac{d^2k_{2m-2}}{(2\pi)^2} \frac{-i}{\alpha - k_{2m-2}^2 - i0}$$

$$\times \int \frac{d^2k_{2m-1}}{(2\pi)^2} \frac{-i}{\alpha - k_{2m-1}^2 - i0} \frac{-i}{\alpha - (k_1 + \cdots + k_{2m-1})^2 - i0}$$

$$- Z_1^2 \alpha^2 \sum_{m=2}^{\infty} \frac{\beta^{4m-4}}{(2m-1)!} \exp\left\{i\beta^2 \Delta_F(0)\right\}$$

$$\times \left(\frac{-i}{\alpha - p^2 - i0}\right)^2 \int \frac{d^2k_1}{(2\pi)^2} \frac{-i}{\alpha - k_1^2 - i0} \cdots$$

$$\times \int \frac{d^2k_{2m-2}}{(2\pi)^2} \frac{-i}{\alpha - k_{2m-2}^2 - i0} \frac{-i}{\alpha - (p - k_1 - \cdots k_{2m-2})^2 - i0},$$

(B.12)

finding

$$G^{(2)}(x,y) =$$

$$= -\alpha^2 Z_1 \frac{\beta^2}{2} \exp\left\{i\frac{\beta^2}{2}\Delta_F(0)\right\}$$

$$\times \iint d^2z_1 d^2z_2 [-i\Delta_F(x-z_1)][-i\Delta_F(y-z_1)][-i\Delta_F(z_1-z_2)]^2$$

$$+ \frac{Z_1^2 \alpha^2}{\beta^2} \exp\left\{i\beta^2 \Delta_F(0)\right\}$$

$$\times \iint d^2z_1 d^2z_2 [-i\Delta_F(x-z_1)] \left(\cosh[(-i\beta^2 \Delta_F(z_1-z_2)] - 1\right)[-i\Delta_F(y-z_1)]$$

$$-\frac{Z_1^2\alpha^2}{\beta^2}\exp\left\{i\beta^2\Delta_F(0)\right\}$$
$$\times \iint d^2z_1 d^2z_2 [-i\Delta_F(x-z_1)]\left(\sinh[(-i\beta^2\Delta_F(z_1-z_2)] + i\beta^2\Delta_F(z_1-z_2)\right)[-i\Delta_F(y-z_2)],$$
(B.13)

and

$$\tilde{G}^{(2)}(p) = i\alpha_{ph}\frac{\beta^2}{8\pi}\left(\frac{-i}{\alpha_{ph}-p^2-i0}\right)^2$$
$$+ \alpha_{ph}^2 \sum_{m=1}^{\infty}\frac{\beta^{4m-2}}{(2m)!}\left(\frac{-i}{\alpha_{ph}-p^2-i0}\right)^2 \int \frac{d^2k_1}{(2\pi)^2}\frac{-i}{\alpha_{ph}-k_1^2-i0}\cdots$$
$$\times \int \frac{d^2k_{2m-2}}{(2\pi)^2}\frac{-i}{\alpha_{ph}-k_{2m-2}^2-i0}$$
$$\times \int \frac{d^2k_{2m-1}}{(2\pi)^2}\frac{-i}{\alpha_{ph}-k_{2m-1}^2-i0}\frac{-i}{\alpha_{ph}-(k_1+\cdots+k_{2m-1})^2-i0}$$
$$- \alpha_{ph}^2 \sum_{m=2}^{\infty}\frac{\beta^{4m-4}}{(2m-1)!}\left(\frac{-i}{\alpha_{ph}-p^2-i0}\right)^2 \int \frac{d^2k_1}{(2\pi)^2}\frac{-i}{\alpha_{ph}-k_1^2-i0}\cdots$$
$$\times \int \frac{d^2k_{2m-2}}{(2\pi)^2}\frac{-i}{\alpha_{ph}-k_{2m-2}^2-i0}\frac{-i}{\alpha_{ph}-(p-k_1-\cdots k_{2m-2})^2-i0},$$
(B.14)

where we have used

$$\int \frac{d^2k}{(2\pi)^2}\left(\frac{-i}{\alpha-k^2-i0}\right)^2 = \frac{1}{4\pi i\alpha},$$
$$\alpha Z_1 \exp\left\{i\frac{\beta^2}{2}\Delta_F(0)\right\} = \alpha_{ph},$$
(B.15)

in the expression above. The correction to order β^2 reads

$$\tilde{G}^{(2,2)}(p) = i\alpha_{ph}\frac{\beta^2}{8\pi}\left(\frac{-i}{\alpha_{ph}-p^2-i0}\right)^2 + \alpha_{ph}^2\frac{\beta^2}{2}\left(\frac{-i}{\alpha_{ph}-p^2-i0}\right)^2 \int \frac{d^2k}{(2\pi)^2}\left(\frac{-i}{\alpha_{ph}-k^2-i0}\right)^2$$
$$= 0,$$
(B.16)

yielding the final result Eqs. (2.72) and (2.73).

B.1.3. The massive sine–Gordon model

Using the same procedure the second order correction to the two–point Green function in α–expansion of the massive sine–Gordon is defined by

$$G^{(2)}_{(m)}(x-y) =$$
$$= G^{(2)}_{(\mu)}(x-y) - \frac{1}{8}m^4(Z_m-1)^2 \iint d^2z_1 d^2z_2 \langle 0|\mathrm{T}(\varphi(x)\varphi(y)\varphi^2(z_1)\varphi^2(z_2))|0\rangle_c$$

B.1. Expansion to second order in α

$$+ \frac{1}{2} m^2 (Z_m - 1) \alpha \sum_{m=2}^{\infty} \frac{(-1)^m}{(2m)!} \beta^{2(m-1)} \iint d^2z_1 d^2z_2 \, \langle 0|T(\varphi(x)\varphi(y)\varphi^2(z_1)\varphi^{2m}(z_2))|0\rangle_c$$

$$+ \frac{1}{2} m^2 (Z_m - 1)(Z_1 - 1) \alpha \sum_{m=1}^{\infty} \frac{(-1)^m}{(2m)!} \beta^{2(m-1)} \iint d^2z_1 d^2z_2 \, \langle 0|T(\varphi(x)\varphi(y)\varphi^2(z_1)\varphi^{2m}(z_2))|0\rangle_c$$

$$= G^{(2)}_{(\mu)}(x-y) - m^4 (Z_m - 1)^2 \iint d^2z_1 d^2z_2 \, [-i\Delta_F(x-z_1)][-i\Delta_F(z_1-z_2)][-i\Delta_F(y-z_2)]$$

$$+ \frac{1}{2} m^2 (Z_m - 1) \alpha$$

$$\times \left[\beta^2 \exp\left\{ i\frac{\beta^2}{2} \Delta_F(0) \right\} \iint d^2z_1 d^2z_2 \, [-i\Delta_F(x-z_1)][-i\Delta_F(y-z_1)][-i\Delta_F(z_1-z_2)]^2 \right.$$

$$\left. - 4 \left(\exp\left\{ i\frac{\beta^2}{2} \Delta_F(0) \right\} - 1 \right) \iint d^2z_1 d^2z_2 \, [-i\Delta_F(x-z_1)][-i\Delta_F(y-z_2)][-i\Delta_F(z_1-z_2)] \right]$$

$$+ \frac{1}{2} m^2 (Z_m - 1)(Z_1 - 1) \alpha$$

$$\times \left[\beta^2 \exp\left\{ i\frac{\beta^2}{2} \Delta_F(0) \right\} \iint d^2z_1 d^2z_2 \, [-i\Delta_F(x-z_1)][-i\Delta_F(y-z_1)][-i\Delta_F(z_1-z_2)]^2 \right.$$

$$\left. - 4 \exp\left\{ i\frac{\beta^2}{2} \Delta_F(0) \right\} \iint d^2z_1 d^2z_2 \, [-i\Delta_F(x-z_1)][-i\Delta_F(y-z_2)][-i\Delta_F(z_1-z_2)] \right]$$

$$= \alpha Z_1 \left[m^2 (Z_m - 1) - \alpha \right] \frac{\beta^2}{2} \exp\left\{ i\frac{\beta^2}{2} \Delta_F(0) \right\}$$

$$\times \iint d^2z_1 d^2z_2 \, [-i\Delta_F(x-z_1)][-i\Delta_F(y-z_1)][-i\Delta_F(z_1-z_2)]^2$$

$$+ \left[-\alpha^2 \left(1 - Z_1 \exp\left\{ i\frac{\beta^2}{2} \Delta_F(0) \right\} \right)^2 - m^4 (Z_m - 1)^2 \right.$$

$$\left. - 2m^2 (Z_m - 1) \alpha Z_1 \exp\left\{ i\frac{\beta^2}{2} \Delta_F(0) \right\} + 2m^2 (Z_m - 1) \alpha \right]$$

$$\times \iint d^2z_1 d^2z_2 \, [-i\Delta_F(x-z_1)][-i\Delta_F(y-z_2)][-i\Delta_F(z_1-z_2)]$$

$$+ \frac{Z_1^2 \alpha^2}{\beta^2} \exp\left\{ i\beta^2 \Delta_F(0) \right\}$$

$$\times \iint d^2z_1 d^2z_2 [-i\Delta_F(x-z_1)] \Big(\cosh[(-i\beta^2 \Delta_F(z_1-z_2)] - 1 \Big) [-i\Delta_F(y-z_1)]$$

$$- \frac{Z_1^2 \alpha^2}{\beta^2} \exp\left\{ i\beta^2 \Delta_F(0) \right\}$$

$$\times \iint d^2z_1 d^2z_2 [-i\Delta_F(x-z_1)] \Big(\sinh[(-i\beta^2 \Delta_F(z_1-z_2)] + i\beta^2 \Delta_F(z_1-z_2) \Big) [-i\Delta_F(y-z_2)], \tag{B.17}$$

where $G^{(2)}_{(\mu)}(x-y)$ is $G^{(2)}(x-y)$ of the massless sine–Gordon model but with the mass μ^2 replacing α. Furthermore, we have used

$$\iint d^2z_1 d^2z_2 \, \langle 0|T(\varphi(x)\varphi(y)\varphi^2(z_1)\varphi^2(z_2))|0\rangle_c =$$
$$= 8 \iint d^2z_1 d^2z_2 \, [-i\Delta_F(x-z_1)][-i\Delta_F(z_1-z_2)][-i\Delta_F(y-z_2)], \tag{B.18}$$

and

$$\sum_{m=A}^{\infty} \frac{(-1)^m \beta^{2(m-1)}}{(2m)!} \iint d^2z_1 d^2z_2 \, \langle 0|\mathrm{T}(\varphi(x)\varphi(y)\varphi^2(z_1)\varphi^{2m}(z_2))|0\rangle_c$$

$$= \sum_{m=A}^{\infty} \frac{(-1)^m \beta^{2(m-1)}}{(2m)!} \Big\{ (1-\delta_{m1}) \frac{(2m)!}{(m-2)! 2^{m-2}}$$

$$\times \iint d^2z_1 d^2z_2 \, [-i\Delta_F(x-z_1)][-i\Delta_F(y-z_1)][-i\Delta_F(z_1-z_2)]^2 [-i\Delta_F(0)]^{m-2}$$

$$+ \underbrace{(\longleftrightarrow)}_{=0}$$

$$+ \frac{(2m)!}{(m-1)! 2^{m-2}} \iint d^2z_1 d^2z_2 \, [-i\Delta_F(x-z_1)][-i\Delta_F(y-z_2)][-i\Delta_F(z_1-z_2)][-i\Delta_F(0)]^{m-1}$$

$$+ \underbrace{(\longleftrightarrow)}_{=\text{factor of 2}} \Big\}$$

$$= \beta^2 \sum_{m=0}^{\infty} \frac{(-1)^m \beta^{2m}}{m! 2^m} [-i\Delta_F(0)]^m \iint d^2z_1 d^2z_2 \, [-i\Delta_F(x-z_1)][-i\Delta_F(y-z_1)][-i\Delta_F(z_1-z_2)]^2$$

$$- 4 \sum_{m=\delta_{A2}}^{\infty} \frac{(-1)^m \beta^{2m}}{m! 2^m} [-i\Delta_F(0)]^m \iint d^2z_1 d^2z_2 \, [-i\Delta_F(x-z_1)][-i\Delta_F(y-z_2)][-i\Delta_F(z_1-z_2)]$$

$$= \beta^2 \exp\Big\{i\frac{\beta^2}{2}\Delta_F(0)\Big\} \iint d^2z_1 d^2z_2 \, [-i\Delta_F(x-z_1)][-i\Delta_F(y-z_1)][-i\Delta_F(z_1-z_2)]^2$$

$$- 4\Big(\exp\Big\{i\frac{\beta^2}{2}\Delta_F(0)\Big\} - \delta_{A2}\Big) \iint d^2z_1 d^2z_2 \, [-i\Delta_F(x-z_1)][-i\Delta_F(y-z_2)][-i\Delta_F(z_1-z_2)].$$

(B.19)

B.1.4. The momentum representation

In the momentum representation we find

$$\tilde{G}^{(2)}_{(m)}(p) =$$

$$= \alpha Z_1 \Big[m^2(Z_m-1) - \alpha\Big]\frac{\beta^2}{2}\exp\Big\{i\frac{\beta^2}{2}\Delta_F(0)\Big\}\Big(\frac{-i}{\mu^2-p^2-i0}\Big)^2 \int \frac{d^2k}{(2\pi)^2}\Big(\frac{-i}{\mu^2-k^2-i0}\Big)^2$$

$$+ \Big[-\alpha^2\Big(1 - Z_1 \exp\Big\{i\frac{\beta^2}{2}\Delta_F(0)\Big\}\Big)^2 - m^4(Z_m-1)^2$$

$$- 2m^2(Z_m-1)\alpha Z_1 \exp\Big\{i\frac{\beta^2}{2}\Delta_F(0)\Big\} + 2m^2(Z_m-1)\alpha\Big]\Big(\frac{-i}{\mu^2-p^2-i0}\Big)^3$$

$$+ Z_1^2 \alpha^2 \sum_{m=1}^{\infty} \frac{\beta^{4m-2}}{(2m)!} \exp\Big\{i\beta^2\Delta_F(0)\Big\}$$

$$\times \Big(\frac{-i}{\mu^2-p^2-i0}\Big)^2 \int \frac{d^2k_1}{(2\pi)^2} \frac{-i}{\mu^2-k_1^2-i0} \cdots \int \frac{d^2k_{2m-2}}{(2\pi)^2} \frac{-i}{\mu^2-k_{2m-2}^2-i0}$$

$$\times \int \frac{d^2k_{2m-1}}{(2\pi)^2} \frac{-i}{\mu^2-k_{2m-1}^2-i0} \frac{-i}{\mu^2-(k_1+\cdots+k_{2m-1})^2-i0}$$

$$- Z_1^2 \alpha^2 \sum_{m=2}^{\infty} \frac{\beta^{4m-4}}{(2m-1)!} \exp\Big\{i\beta^2\Delta_F(0)\Big\}\Big(\frac{-i}{\mu^2-p^2-i0}\Big)^2 \int \frac{d^2k_1}{(2\pi)^2} \frac{-i}{\mu^2-k_1^2-i0} \cdots$$

$$\times \int \frac{d^2 k_{2m-2}}{(2\pi)^2} \frac{-i}{\mu^2 - k_{2m-2}^2 - i0} \frac{-i}{\mu^2 - (p - k_1 - \cdots k_{2m-2})^2 - i0}$$

$$= \alpha Z_1 \left[m^2(Z_m - 1) - \alpha \right] \frac{\beta^2}{2} \exp\left\{ i\frac{\beta^2}{2} \Delta_F(0) \right\} \left(\frac{-i}{m_{ph}^2 - p^2 - i0} \right)^2 \int \frac{d^2 k}{(2\pi)^2} \left(\frac{-i}{m_{ph}^2 - k^2 - i0} \right)^2$$

$$+ Z_1^2 \alpha^2 \sum_{m=1}^{\infty} \frac{\beta^{4m-2}}{(2m)!} \exp\left\{ i\beta^2 \Delta_F(0) \right\}$$

$$\times \left(\frac{-i}{m_{ph}^2 - p^2 - i0} \right)^2 \int \frac{d^2 k_1}{(2\pi)^2} \frac{-i}{m_{ph}^2 - k_1^2 - i0} \cdots \int \frac{d^2 k_{2m-2}}{(2\pi)^2} \frac{-i}{m_{ph}^2 - k_{2m-2}^2 - i0}$$

$$\times \int \frac{d^2 k_{2m-1}}{(2\pi)^2} \frac{-i}{m_{ph}^2 - k_{2m-1}^2 - i0} \frac{-i}{m_{ph}^2 - (k_1 + \cdots + k_{2m-1})^2 - i0}$$

$$- Z_1^2 \alpha^2 \sum_{m=2}^{\infty} \frac{\beta^{4m-4}}{(2m-1)!} \exp\left\{ i\beta^2 \Delta_F(0) \right\} \left(\frac{-i}{m_{ph}^2 - p^2 - i0} \right)^2 \int \frac{d^2 k_1}{(2\pi)^2} \frac{-i}{m_{ph}^2 - k_1^2 - i0} \cdots$$

$$\times \int \frac{d^2 k_{2m-2}}{(2\pi)^2} \frac{-i}{m_{ph}^2 - k_{2m-2}^2 - i0} \frac{-i}{m_{ph}^2 - (p - k_1 - \cdots k_{2m-2})^2 - i0}. \tag{B.20}$$

B.2. Exact solutions concerning Gaussian quantum corrections around a soliton

First the relation Eq. (2.92) is derived. Using (see Ref. [83])

$$\cos 4z = 8 \cos^4 z - 8 \cos^2 z + 1, \tag{B.21}$$

and the relation (see Ref. [84])

$$\arctan x = \arccos \frac{1}{\sqrt{1+x^2}} \qquad x \geq 0,$$
$$= -\arccos \frac{1}{\sqrt{1+x^2}} \qquad x < 0, \tag{B.22}$$

we find for

$$\cos(4 \arctan \exp(\sqrt{\alpha_0} \rho)) = 1 + 8 \left(\cos^4 \arctan \exp(\sqrt{\alpha_0} \rho) - \cos^2 \arctan \exp(\sqrt{\alpha_0} \rho) \right)$$

$$= 1 + 8 \left(\frac{1}{[1 + \exp(2\sqrt{\alpha_0} \rho)]^2} - \frac{1}{1 + \exp(2\sqrt{\alpha_0} \rho)} \right)$$

$$= 1 - 8 \frac{\exp(2\sqrt{\alpha_0} \rho)}{[1 + \exp(2\sqrt{\alpha_0} \rho)]^2}$$

$$= 1 - 8 \frac{1}{[\exp(-\sqrt{\alpha_0} \rho) + \exp(\sqrt{\alpha_0} \rho)]^2}$$

$$= 1 - \frac{2}{\cosh^2(\sqrt{\alpha_0} \rho)}, \tag{B.23}$$

arriving at Eq. (2.92).

Now to the solution of Eq. (2.101). Introducing $\eta = \tanh(\sqrt{\alpha_0}\rho)$ we reduce Eq. (2.101) to (see Ref. [81])

$$\frac{d}{d\eta}\left[(1-\eta^2)\frac{d\psi(\eta)}{d\eta}\right] + \left[s(s+1) - \frac{\epsilon^2}{1-\eta^2}\right]\psi(\eta) = 0, \qquad (B.24)$$

with $s(s+1) = 2$ and $\epsilon^2 = -k^2/\alpha_0 = 1 - \omega^2/\alpha_0$. Substituting $\psi(\eta) = (1-\eta^2)^{\epsilon/2} w(\eta)$ and introducing $u = (1-\eta)/2$ we find

$$u(1-u)w'' + (\epsilon+1)(1-2u)\,w' - (\epsilon-s)(\epsilon+s+1)\,w = 0. \qquad (B.25)$$

The solution of this equation reads (see Ref. [82, 83])

$$w(\eta) = w^{(1)}(\eta) + w^{(2)}(\eta)$$
$$= C_1 F\left(\epsilon - s, \epsilon + s + 1; \epsilon + 1; \frac{1-\eta}{2}\right) + C_2 \left(\frac{1-\eta}{2}\right)^{-\epsilon} F\left(-s, s+1; 1-\epsilon; \frac{1-\eta}{2}\right), \qquad (B.26)$$

with the integration constants C_1 and C_2.

The parameter s takes the values $s = -2, +1$, which solve the equation $s(s+1) = 2$. The hypergeometric functions coincide for both cases (see Ref. [82]). Therefore, we set $s = 1$ and find

$$w(\eta) = w^{(1)}(\eta) + w^{(2)}(\eta)$$
$$= C_1 F\left(\epsilon - 1, \epsilon + 2; \epsilon + 1; \frac{1-\eta}{2}\right) + C_2 \left(\frac{1-\eta}{2}\right)^{-\epsilon} F\left(-1, 2; 1-\epsilon; \frac{1-\eta}{2}\right), \qquad (B.27)$$

for the solution of $w^{(2)}(\eta)$ we find (see Ref. [82])

$$w^{(2)}(\eta) = C_2 \left(\frac{1-\eta}{2}\right)^{-\epsilon} F\left(2, -1; 1-\epsilon; \frac{1-\eta}{2}\right)$$
$$= C_2 \left(\frac{1-\eta}{2}\right)^{-\epsilon} F\left(-1, 2; 1-\epsilon; \frac{1-\eta}{2}\right)$$
$$= C_2 \left(\frac{1-\eta}{2}\right)^{-\epsilon} \frac{\eta-\epsilon}{1-\epsilon}. \qquad (B.28)$$

Using (see Refs. [83, 82])

$$F\left(a, b; c; \frac{1-\eta}{2}\right) = \frac{\Gamma(c)\Gamma(c-a-b)}{\Gamma(c-a)\Gamma(c-b)} F\left(a, b; a+b-c+1; \frac{1+\eta}{2}\right)$$
$$+ \left(\frac{1+\eta}{2}\right)^{c-a-b} \frac{\Gamma(c)\Gamma(a+b-c)}{\Gamma(a)\Gamma(b)} F\left(c-a, c-b; c-a-b+1; \frac{1+\eta}{2}\right), \qquad (B.29)$$

we can solve $w^{(1)}(\eta)$

$$w^{(1)}(\eta) = C_1 F\left(\epsilon - 1, \epsilon + 2; \epsilon + 1; \frac{1-\eta}{2}\right)$$

B.2. Exact solutions concerning Gaussian quantum corrections around a soliton

$$
\begin{aligned}
&= C_1 \frac{\Gamma(\epsilon+1)\Gamma(-\epsilon)}{\Gamma(-1)} F\left(\epsilon-1, \epsilon+2; \epsilon+1; \frac{1+\eta}{2}\right) \\
&\quad + C_1 \frac{\Gamma(\epsilon+1)\Gamma(\epsilon)}{\Gamma(\epsilon-1)\Gamma(\epsilon+2)} \left(\frac{1+\eta}{2}\right)^{-\epsilon} F\left(2,-1; 1-\epsilon; \frac{1+\eta}{2}\right) \\
&= C_1 \left(\frac{1+\eta}{2}\right)^{-\epsilon} \frac{\epsilon+\eta}{1+\epsilon}. \quad (B.30)
\end{aligned}
$$

For the field $\psi(\eta)$ we obtain

$$\psi(\eta) = C_1 \, 2^{\epsilon} \left(\frac{1-\eta}{1+\eta}\right)^{\epsilon/2} \frac{\epsilon+\eta}{1+\epsilon} + C_2 \, 2^{\epsilon} \left(\frac{1+\eta}{1-\eta}\right)^{\epsilon/2} \frac{\eta-\epsilon}{1-\epsilon}, \quad (B.31)$$

respectively

$$\psi(\rho) = C_1 \left(\epsilon + \tanh(\sqrt{\alpha_0}\rho)\right) e^{-\epsilon\sqrt{\alpha_0}\rho} + C_2 \left(-\epsilon + \tanh(\sqrt{\alpha_0}\rho)\right) e^{+\epsilon\sqrt{\alpha_0}\rho}, \quad (B.32)$$

with redefined integration constants. For $\epsilon = \pm 1$ we find a discrete mode and a continuum for $\epsilon = \pm ik/\sqrt{\alpha_0}$. So we have

$$
\begin{aligned}
\psi_b(\rho) &= \sqrt{\frac{\sqrt{\alpha_0}}{2}} \, \frac{1}{\cosh(\sqrt{\alpha_0}\rho)}, \\
\psi_k(\rho) &= \frac{i}{\sqrt{2\pi}} \frac{-ik + \sqrt{\alpha_0} \tanh(\sqrt{\alpha_0}\rho)}{\sqrt{k^2 + \alpha_0}} e^{+ik\rho},
\end{aligned}
\quad (B.33)
$$

respectively

$$
\begin{aligned}
\phi_{\omega b}(\tau, \rho) &= \frac{1}{\sqrt{2\pi}} \sqrt{\frac{\sqrt{\alpha_0}}{2}} \frac{1}{\cosh(\sqrt{\alpha_0}\rho)} e^{-i\omega\tau}, \\
\phi_{\omega k}(\tau, \rho) &= \frac{i}{2\pi} \frac{-ik + \sqrt{\alpha_0} \tanh(\sqrt{\alpha_0}\rho)}{\sqrt{k^2 + \alpha_0}} e^{-i\omega\tau + ik\rho}.
\end{aligned}
\quad (B.34)
$$

The norm of these functions is

$$
\begin{aligned}
\int_{-\infty}^{+\infty} d\rho \, \psi_{k'}^*(\rho)\psi_k(\rho) &= \frac{1}{k'^2 - k^2} \left(\psi_{k'}^*(\rho) \frac{d}{d\rho}\psi_k(\rho) - \psi_k(\rho) \frac{d}{d\rho}\psi_{k'}^*(\rho)\right)\bigg|_{-\infty}^{+\infty} \\
&= \delta(k' - k), \\
\int_{-\infty}^{+\infty} d\rho \, |\psi_b(\rho)|^2 &= 1,
\end{aligned}
\quad (B.35)
$$

respectively

$$
\begin{aligned}
\int_{-\infty}^{+\infty}\int_{-\infty}^{+\infty} d\tau \, \phi_{\omega' b}^*(\tau, \rho)\phi_{\omega b}(\tau, \rho) &= \delta(\omega' - \omega), \\
\int_{-\infty}^{+\infty}\int_{-\infty}^{+\infty} d\tau d\rho \, \phi_{\omega' k'}^*(\tau, \rho)\phi_{\omega k}(\tau, \rho) &= \delta(\omega' - \omega)\delta(k' - k),
\end{aligned}
\quad (B.36)
$$

where the function $\psi_b(\rho)$ has eigenvalue $\omega = 0$ and corresponds to a bound state (see

Ref. [76]). The following completeness condition holds

$$\int_{-\infty}^{+\infty} dk\, \psi_k^*(\rho')\psi_k(\rho) + \psi_b(\rho')\psi_b(\rho) = \delta(\rho' - \rho), \tag{B.37}$$

where we have used

$$\begin{aligned}
\int_{-\infty}^{+\infty} &dk\, \psi_k^*(\rho')\psi_k(\rho) + \psi_b(\rho')\psi_b(\rho) = \\
&= \int_{-\infty}^{+\infty} \frac{dk}{2\pi} e^{ik(\rho - \rho')} \\
&+ \sqrt{\alpha_0}\,[\tanh(\sqrt{\alpha_0}\rho) - \tanh(\sqrt{\alpha_0}\rho')]\int_{-\infty}^{+\infty} \frac{dk}{2\pi} \frac{ik}{k^2 + \alpha_0} e^{ik(\rho - \rho')} \\
&+ \alpha_0\,[\tanh(\sqrt{\alpha_0}\rho')\tanh(\sqrt{\alpha_0}\rho) - 1]\int_{-\infty}^{+\infty} \frac{dk}{2\pi} \frac{1}{k^2 + \alpha_0} e^{ik(\rho - \rho')} \\
&+ \frac{\sqrt{\alpha_0}}{2} \frac{1}{\cosh(\sqrt{\alpha_0}\rho')}\frac{1}{\cosh(\sqrt{\alpha_0}\rho)} \\
&= \delta(\rho' - \rho) - \frac{\sqrt{\alpha_0}}{2} \frac{1}{\cosh(\sqrt{\alpha_0}\rho')}\frac{1}{\cosh(\sqrt{\alpha_0}\rho)} e^{-\sqrt{\alpha_0}|\rho - \rho'|} \\
&\quad \times [\varepsilon(\rho - \rho')\sinh(\sqrt{\alpha_0}(\rho - \rho')) + \cosh(\sqrt{\alpha_0}(\rho - \rho'))] \\
&+ \frac{\sqrt{\alpha_0}}{2} \frac{1}{\cosh(\sqrt{\alpha_0}\rho')}\frac{1}{\cosh(\sqrt{\alpha_0}\rho)} \\
&= \delta(\rho' - \rho),
\end{aligned} \tag{B.38}$$

the relation

$$e^{+\sqrt{\alpha_0}|\rho - \rho'|} = \varepsilon(\rho - \rho')\sinh(\sqrt{\alpha_0}((\rho - \rho')) + \cosh(\sqrt{\alpha_0}(\rho - \rho')), \tag{B.39}$$

and the sign–function $\varepsilon(\rho - \rho')$.

Bibliography

[1] W. Thirring, Ann. Phys. (N.Y.) **3**, 91 (1958).

[2] V. Glaser, Nuovo Cim. **9**, 990 (1958).

[3] F. Scarf, Phys. Rev. **117**, 868 (1960).

[4] T. Pradhan, Nucl. Phys. **9**, 124 (1958-59).

[5] K. Johnson, Nuovo Cim. **20**, 773 (1961).

[6] C. Sommerfield, Ann. of Phys. **26**, 1 (1964).

[7] C. R. Hagen, Nuovo Cim. B **51**, 169 (1967).

[8] J. Schwinger, Phys. Rev. **128**, 2425 (1962).

[9] B. Klaiber, in *LECTURES IN THEORETICAL PHYSICS*, Lectures delivered at the Summer Institute for Theoretical Physics, University of Colorado, Boulder, 1967, edited by A. Barut and W. Brittin, Gordon and Breach, New York, 1968, Vol. X, part A, pp.141–176.

[10] M. Faber and A. N. Ivanov, hep–th/0112183, 2001.

[11] R. Jackiw, Phys. Rev. D **3**, 2005 (1971).

[12] F. L. Scarf and J. Wess, Nuovo Cim. **26**, 150 (1962).

[13] K. Furuya, Re. E. Gamboa Saravi and F. A. Schaposnik, Nucl. Phys. B **208**, 159 (1982).

[14] C. M. Naón, Phys. Rev. D **31**, 2035 (1985).

[15] R. Roskies and F. A. Schaposnik, Phys. Rev. D **23**, 558 (1981).

[16] R. E. Gamboa Saravi, F. A. Schaposnik, and J. E. Solomin, Nucl. Phys. B **153**, 112 (1979).

[17] R. E. Gamboa Savari, M. A. Muschetti, F. A. Schaposnik, and J. E. Solomin, Ann. of Phys. (N.Y.) **157**, 360 (1984).

[18] O. Alvarez, Nucl. Phys. B **238**, 61 (1984).

[19] H. Dorn, Phys. Lett. B **167**, 86 (1986).

[20] J. C. Collins, in *RENORMALIZATION, An Introduction to Renormalisation, the Renormalisation Group, and the Operator–Product Expansion*, Cambridge University Press, Cambridge, 1984.

[21] J. Schwinger, Phys. Rev. Lett. **3**, 296 (1959).

[22] K. Harada, H. Kubota, and I. Tsutsui, Phys. Lett. B **173**, 77 (1986).

[23] R. Jackiw and R. Rajaraman, Phys. Rev. Lett. **54**, 1219 (1985).

[24] G. A. Christos, Z. Phys. C **18**, 155 (1983); Erratum Z. Phys. C **20**, 186 (1983).

[25] A. Smailagic and R. E. Gamboa–Saravi, Phys. Lett. B **192**, 145 (1987).

[26] R. Banerjee, Z. Phys. C **25**, 251 (1984).

[27] T. Ikehashi, Phys. Lett. B **313**, 103 (1993).

[28] A. H. Mueller and T. L. Trueman, Phys. Rev. D **4**, 1635 (1971).

[29] M. Gomes and J. H. Lowenstein, Nucl. Phys. B **45**, 252 (1972).

[30] S. Coleman, Phys. Rev. D **11**, 2088 (1975).

[31] J. M. Kosterlitz and D. J. Thouless, J. Phys. C **6**, 118 (1973).

[32] J. M. Kosterlitz, J. Phys. C **7**, 1046 (1974).

[33] J. V. Jose, L. P. Kadanoff, S. Kirkpatrick, and D. R. Nelson, Phys. Rev. B **16**, 1217 (1977).

[34] P. B. Wiegmann, J. Phys. C **11**, 1583 (1978).

[35] S. Samuel, Phys. Rev. D **18**, 1916 (1978).

[36] D. Amit, Y. Y. Goldschmidt, and G. Grinstein, J. Phys. A **13**, 585 (1980).

[37] B. Nienhuis, in *Phase transitions and critical phenomena*, edited by C. Domb and J. L. Lebowitz, Academic, London, Vol. 11, pp.1–53, 1987.

[38] K. Huang and J. Polonyi, Int. Mod. Phys. A **6**, 409 (1991).

[39] R. J. Creswick, H. A. Farach, and C. P. Poole, Jr., in *Introduction to RG Methods in Physics*, Wiley, New York, 1998.

[40] Zs. Gulácsi and M. Gulácsi, Adv. Phys. **47**, 1 (1998).

[41] I. Nándori, J. Polonyi, and K. Sailer, Phys. Rev. D **63**, 045022 (2001).

[42] I. Nándori, J. Polonyi, and K. Sailer, Philos. Mag. B **81**, 1615 (2001).

[43] G. von Gersdorf and C. Wetterich, Phys. Rev. B **64**, 054513 (2001).

[44] I. Nándori, K. Sailer, U. D. Jentschura, and G. Soff, J. Phys. G **28**, 607 (2002).

[45] H. A. Fertig and K. Majumdar, cond–mat/0302012.

[46] M. Faber and A. N. Ivanov, J. Phys. A **36**, 7839 (2003) and references therein.

[47] I. Nándori, K. Sailer, U. D. Jentschura, and G. Soff, Phys. Rev. D **69**, 025004 (2004).

[48] H. Bozkaya, M. Faber, A. N. Ivanov, M. Pitschmann, J. Phys. A: Math. Gen. **39**, 2177 (2006).

[49] R. F. Streater and A. S. Wightman, in *PCT, spin and statistics*, Princeton University Press, Princeton and Oxford, Third Edition, 1980.

[50] S. Coleman, Comm. Math. Phys. **31**, 259 (1973).

[51] R. Rajaraman, in *Solitons and Instantons*, North–Holland, 1982.

[52] N. J. Zabusky and M. D. Kruskal, Phys. Rev. Lett. **15**, 240 (1965).

[53] T. H. R. Skyrme, Proc. R. Soc. **A247**, 260 (1958).

[54] J. K. Perring and T. H. R. Skyrme, Nucl. Phys. **31**, 550 (1962).

[55] J. Frenkel and T. Kontrova, J. Phys. USSR **1**, 137 (1939).

[56] A. Seeger, H. Donth and A. Kochendorfer, Z. Physik **134**, 173 (1953).

[57] S. L. McCall and E. L. Hahn, Phys. Rev. Lett. **18**, 908 (1967).

[58] M. Faber and A. N. Ivanov, Eur. Phys. J. **C20**, 723 (2001).

[59] S. Coleman, Annals Phys. **101**, 239 (1976).

[60] A. Wereszczynski, Acta Phys. Polon. **B31**, 1885 (2000).

[61] A. C. Scott, F. Y. F. Chiu and D. W. Mclaughlin, Proc. I.E.E.E. **61**, 1443.

[62] A. Barone, F. Esposito, C. J. Magee and A. C. Scott, Riv. Nuovo Cim. **1**, 227 (1971).

[63] C. S. Gardner, J. M. Greene, M. D. Kruskal and R. M. Miura, Phys. Rev. Lett. **19**, 1095 (1967).

[64] G. L. Lamb, Jr., Rev. Mod. Phys. **43**, 99 (1971).

[65] T. Barnard, Phys. Rev. **A7**, 373 (1973).

[66] C. Itzykson and J.-B. Zuber, in *Quantum field theory*, McGraw–Hill, New York, 1980.

[67] M. E. Peskin and D. V. Schroeder, in *An Introduction to Quantum Field Theory*, The Advanced Book Program, Westview Press, 1995.

[68] S. Weinberg, in *The quantum theory of fields*, Vol. II *Modern Applications*, Cambridge University Press, 1996.

[69] S. Weinberg, Phys. Rev. **118**, 838 (1960).

[70] J. D. Bjorken and S. D. Drell, *Relativistic Quantum Fields*, McGraw-Hill, 1965.

[71] R. F. Dashen, B. Hasslacher, and A. Neveu, Phys. Rev. D **10**, 4130 (1974).

[72] R. F. Dashen, B. Hasslacher, and A. Neveu, Phys. Rev. D **11**, 3424 (1975).

[73] L. D. Faddeev and V. E. Korepin, Phys. Rep. **42**, 1 (1978).

[74] N. N. Bogoliubov and D. V. Shirkov, in *Introduction to the quantum theory of quantized fields*, Interscience Publishers, Inc., New York, 1959.

[75] Alexander B. Zamolodchikov and Alexey B. Zamolodchikov, Ann. of Phys. **120**, 253 (1979).

[76] J. Rubinstein, J. of Math. Phys. **11**, 258 (1970).

[77] N. N. Lebedev, I. P. Slalskaya, and Y. S. Uflyand, in *Problems of Mathematical Physics*, Prentice–Hall, Inc., Englewood Cliffs, New Jersey, 1965, pp.55–102.

[78] (see [67] p.286)

[79] A. Rebhan and P. van Nieuwenhuizen, Nucl. Phys. B **508**, 449 (1997).

[80] L. Dolan and R. Jackiw, Phys. Rev. D **9**, 3320 (1974).

[81] L. D. Landau and E. M. Lifshitz, in *QUANTUM MECHANICS*, Pergamon Press, New York, 1965.

[82] N. N. Lebedew, in *SPEZIELLE FUNKTIONEN UND IHRE ANWENDUNG*, Wissenschaftsverlag, Bibliographisches Institut, Mannheim, 1973.

[83] *HANDBOOK OF MATHEMATICAL FUNCTIONS, with Formulas, Graphs, and Mathematical Tables*, ed. by M. Abramowitz and I. E. Stegun, U.S. Department of Commerce, National Bureau of Standards, Applied Mathematics Series • 55, 1972.

[84] K. Bosch, in *Mathematik-Taschenbuch*, R. Oldenbourg Verlag, 1993.

Die VDM Verlagsservicegesellschaft sucht für wissenschaftliche Verlage abgeschlossene und herausragende

Dissertationen, Habilitationen, Diplomarbeiten, Master Theses, Magisterarbeiten usw.

für die kostenlose Publikation als Fachbuch.

Sie verfügen über eine Arbeit, die hohen inhaltlichen und formalen Ansprüchen genügt, und haben Interesse an einer honorarvergüteten Publikation?

Dann senden Sie bitte erste Informationen über sich und Ihre Arbeit per Email an *info@vdm-vsg.de*.

Sie erhalten kurzfristig unser Feedback!

VDM Verlagsservicegesellschaft mbH
Dudweiler Landstr. 99
D - 66123 Saarbrücken

Telefon +49 681 3720 174
Fax +49 681 3720 1749

www.vdm-vsg.de

Die VDM Verlagsservicegesellschaft mbH vertritt

Printed by Books on Demand GmbH, Norderstedt / Germany